[Document E.]

BY THE SENATE,
February 12, 1860.

3,000 copies ordered to be printed.

FIRST REPORT

OF

PHILIP T. TYSON,

STATE AGRICULTURAL CHEMIST,

TO THE

HOUSE OF DELEGATES OF MARYLAND,

JANUARY, 1860.

Baltimore, January 9, 1860.

To the Hon. E. G. KILBOURN,

Speaker of the House of Delegates.

Sir:—

I have the honor to transmit, herewith, my First Report, with the accompanying Maps.

Very respectfully, yours,

PHILIP T. TYSON,

State Agricultural Chemist.

CONTENTS OF THE REPORT

PAGE.

PREFACE........ 5
Necessity for Improvement in Agriculture........ 5
Production has not kept pace with population in the Atlantic States....... 5
Increasing Desire for Knowledge........ 6
Scientific Knowledge required by the Farmer........ 7
Industrial character of a country depends upon its Geological character... 7
Surveys in Maryland necessary........ 8
Compilation of a Large Map........ 9
"Geological Illustrations" accompanying this Report........ 10

CHAPTER I.

MINERALS COMPOSING THE ROCKS OF THIS STATE.

Quartz........ 13
Felspar........ 14
Mica........ 15
Talc and Steatite........ 16
Serpentine........ 16
Chlorite........ 17
Augite and Pyroxene........ 17
Hornblende........ 17
Epidote........ 18
Carbonate of Lime........ 19
Carbonate of Lime, Magnesian........ 19
Oxides of Iron........ 19
Oxide of Manganese........ 20
Sulphur and Sulphates........ 20
Natural Supplies of Gypsum to the Soil........ 20
Phosphoric Acid and Phosphates........ 20
Chlorine........ 21

CHAPTER II.

MINERAL CHARACTERS OF ROCKS.

1. ROCKS OF IGNEOUS ORIGIN........ 22
   Granite........ 22
   Syenite........ 23

Quartz Rock 23
Porphyry 23
Amygdaloid 24
Trap, Amphibolite, or Hornblend Rock 24
Serpentine Rock 24
2. ROCKS OF AQUEOUS ORIGIN 25
Chemical Deposites 25
Limestone 25
Limestone, Magnesian or Dolomite 25
3. SEDIMENTARY DEPOSITS 25
Sandstone 25
Conglomerate 26
Breccia 26
Clay Slates 27
Shales 27
Clays 27
4. METAMORPHIC ROCKS 28
Gneiss 28
Mica-Slate 29
Hornblende Slate 29
Talcose Slate 29
Chlorite Slate 30
Quartzite 30
Granular Limestone 30
Magnesian or Dolomite 30

CHAPTER III.

CONSIDERS THE ABOVE ROCKS AS GROUPED INTO GEOGRAPHICAL FORMATIONS, AND ALSO THEIR GEOGRAPHIC DISTRIBUTION IN MARYLAND.

Nomenclature of the Formations 30
Formation No. 5, including Gneiss, Mica-Slate and Hornblende Slate 32
Formation 6 and 7, consisting of Talc Slates and Roof Slates 33
" 8 and 9, Pottsdam Sandstone and Slate 34
" 10, Limestone 35
" 11, Metamorphic Limestones 35
" 12, Argillaceous Limestone and Slate 36
" 13, Gray and Red Sandstone 36
" 14a, Clinton Group of Shales and Iron Ore 37
" 14b, Shales and Limestones, (Salt Group of N. Y.) 37
" 15a, Limestone, (Meridian of Pennsylvania Reports.) 38
" 15b, Oriscany Sandstone of N. Y 38
" 16a, Shales 38
" 16b, Slates, Sandstones, Flag Stones and Shales 39
" 17b, Old Red Sandstone and Shales 39
" 18a, Shales and Sandstones 40
" 18b, Shales and Limestones 40

Formation 19, Coal Formations........ 40
" 20, New Red Sandstone........ 41
" 21, Cretaceous Group........ 41
" Iron Ore Clays........ 42
" Green Sand Marl........ 42
" 23, Tertiary Group........ 42
" Shell Marls........ 42
" 24, Post Tertiary........ 44

CHAPTER IV.

CHEMICAL AND PHYSICAL GEOLOGY, in which the causes of the destruction of rocks and the production of soils are explained...... 45
Cause of the varieties of soil........ 49

CHAPTER V.

CONSTITUENTS OF PLANTS.

Organic Elements of Plants........ 52
Mineral " " "........ 53

CHEMICAL COMPOSITION OF THE ASHES OF PLANTS.

Indian Corn........ 54
Wheat........ 54
Rye........ 55
Oats........ 55
Potatoes........ 56
Tobacco........ 56
Red Clover........ 56
Timothy Grass........ 57
Further research needed in this connection........ 58

CHAPTER VI.

OF SOILS AND THE CAUSES OF THEIR EXHAUSTION.

Varieties of soils in Maryland........ 59
Liebig's "Mineral Theory"........ 60
Experiments with Artificial Soils........ 62
Soils long cultivated are deficient in Lime, Phosphoric, Acid and humus........ 63

CHAPTER VII.

IMPROVEMENT OF SOILS.

Analysis of Soil........ 64
Our expectations in this respect not realized........ 64
Views of Prof. Anderson thereon........ 64
Difficulties connected with the subject........ 65

Lime and Phosphoric Acid must be applied 66
The rotation should be such as to maintain a full supply of humus or vegetable matter 67
Causes of Short Crops 67
Action of Lime upon Soils 68
Lime promotes the solubility of mineral and vegetable matters 68
Experiments in proof of this 68
Injudicious use of Lime tends to exhaust the Soil 68

CHAPTER VIII.

LIMESTONES OF MARYLAND.

LIMESTONES in Harford, Baltimore and Howard Counties 69
" " Frederick and Carroll 70
" " Washington County 70
" " Allegany County 71
Precautions in selecting Samples for Analysis 71
A minute survey of these Limestone Districts deemed necessary to protect the farmer 71
Composition of Limestones from different localities 72
Amount of Impurities, (Sand, &c.,) in each 72
Table showing the weights of pure lime and impurities in one ton of each kind of stone 73
Use of Magnesian Lime 74
First use of Lime in this State as Manure 74
Its use extended at first slowly 74
Difficulty which formerly existed in inducing those owning "limestone lands" to use lime as manure 74
Oyster Shell Lime 75
" " " weight of per bushel 76

CHAPTER IX.

MARL.

Green Sand or Jersey Marl 76
Its occurrence in Cecil and Kent County 76
" " " Anne Arundel, Prince George's and Charles County 79
Use of, in New Jersey 79
Composition of 80
SHELL MARL 81
" " on the Eastern Shore 81
" " " " Western Shore 81
" " Proportion of Lime in 81
Analysis of Phosphatic Shell Marl and comparison with phosphatic guanos 82
Comparative value of Lime and Marl, as tested by Judge W. B. Carmichael 83

How frequently should Marl be applied........ 84
The quantity per acre........ 84
By whom and when first used in Maryland........ 84
Good effects of the use of Marl on the Eastern Shore........ 84
Little used on the Western Shore........ 85
Cost of digging and freight on Marl........ 85
Marl and lime compared........ 86
Fresh water marl in Washington county........ 88

CHAPTER X.

GUANO.

When first used in Maryland........ 89
Different kinds of guano........ 89
Analysis by Dr. Piggott as follows:........ 90
Brown Mexican........ 91
Patagonian........ 92
Soldanha........ 92
Jarvis Island........ 93
White Mexican........ 93
Colombian........ 94
Sombrero........ 96
(" analyzed by Prof. Morfit)........ 96
Nevassa........ 97
Testigoes........ 97
El Monita........ 98
Peruvian, ammonia contained in California........ 99
(Peruvian, by Prof. Johnson)........ 100
CALIFORNIA GUANO (Elide Island)........ 101
SOMBRERO DO. 100
Adulteration of Guanos........101, 102

CHAPTER XI.

BONES.

BONES FIRST USED AS A MANURE IN GERMANY........ 103
First use of in Maryland........ 103
Lasting effect of bones........ 104
Composition of........104—105
Effect of boiling........ 105
Value of their constituents........ 105
Modes of preparing........ 106
Use of dry ground bones........ 106
Dissolving in sulphuric acid........ 107
Proportions of acid used........ 107
Cost of dissolved bones........ 108
Precautions to be observed in using the acid........ 109
Putrified or "fermented" recommended........ 109

Mode of preparing the latter and of their prompt action upon the crop.... 109
Advantage of the use of bones.... 111
BONE BLACK OR ANIMAL CHARCOAL.... 111
CRACKLINS OR GREAVES.... 111

CHAPTER XII.

MISCELLANEOUS MANURES.

MARSH MUCK AND PEAT.... 112
Abundant in the tide-water counties.... 113
Properties of and modes of using.... 113
SEDIMENT FROM FRESH WATER PONDS.... 114
LIGNITE, USE OF AS A MANURE.... 115
Use of as a destroyer of insects.... 115
Abounds in Maryland.... 115
Experiments with it suggested.... 115
NIGHT SOIL AND POUDRETTE.... 116
Value of.... 117
Sugar refiners scum.... 117
Cost and value of it as a manure.... 117

CHAPTER XIII.

ASHES AND OTHER MANURES.

WOOD ASHES, composition of.... 118
Spent or leached ashes.... 118
COAL ASHES, composition of.... 119
GYPSUM, action of in soils.... 119
Effects produced by.... 120
SOOT FROM CHIMNEYS.... 121
Composition of wood soot.... 121
Mode of using.... 121
Coal soot.... 121
ALKALINE SALTS AS MANURES.... 122
COMMON SALT AS A MANURE.... 122
Utility of.... 122

CHAPTER XIV.

BARN YARD MANURE

Waste of permitted by farmers.... 123
Boussingault's remarks thereon.... 124
Chemistry of the dung hill.... 125
Mode of managing recommended.... 125
Use of plaster therewith.... 126

CHAPTER XV.

ARTIFICIAL MANURES OR FERTILIZERS.

MANIPULATED GUANOS.. 127
SUPERPHOSPHATE OF LIME, COMPOSITION OF WHEN PURE.... 128
FERTILIZERS uncertainty of many of.. 128
" money value of their constituents................................. 129
"National Fertilizer" 129
Mape's Superphosphate...... 130
Degeneration of.. 130
De Burgs superphosphate.. 130
Degeneration of... 130
Coe's superphosphate value of... 130
Rhodes do. do. 130
Jourdan's do. degeneration of.. 130
Money value of in 1857 and 1858................................ 131
Suggestions for protecting our farmers from frauds in fertilizers and adulterated guanos and other manures.................................. 132

CHAPTER XVI.

OBSERVATIONS UPON THE PRESENT STATE OF AGRICULTURE IN MARYLAND, WITH SUGGESTIONS FOR ITS IMPROVEMENT.

Exhaustion of the soil in former days,..................................... 133
Causes of its present improving condition,.. 133
Analysis of Soils,.. 135
Difficulty of drawing correct conclusions from analysis,............. 135
Instance of a remarkable tobacco soil from the Island of Cuba,..... 135
Rotation of Crops,... 135
Effect of green sand marl in improving the quality of potatoes,..... 141
MODES OF APPLYING MANURE,.. 141
1. Lime,.. 141
2. Marl, ... 142
3. Bones,.. 143
4. Guano, ... 143
5. Stable Manure,................. 143
Conclusion,... 144

CONTENTS OF THE APPENDIX.

MINERAL RESOURCES OF MARYLAND,........ 1
MARBLE, in Baltimore County,........ 1
" Carroll " 2
" Frederick " 2
" Washington " 3
" Allegany " 4
HYDRAULIC CEMENT OR WATER LIME........ 4
In Washington County,........ 4
" Allegany " 4
BUILDING STONE........ 5
Granite in Anne Arundel County,........ 5
" " Cecil County,........ 5
Sandstone in Montgomery County,........ 5
" " Frederick " 5
" westward of Frederick county,........ 5
FLAGSTONES, from New York,........ 5
" " Pennsylvania,........ 5
" in Carroll County,........ 5
" " Washington County,........ 6
" " Allegany " 6
ROOFING SLATES in Harford " 6
Prices of Slates in " " 6
In Montgomery " 7
" Frederick " 7
Prices of Slates at Ijamsville........ 7
Importance of the Slates,........ 7
Suggestions for increasing the trade therein,........ 7
CLAY........ 8
Kaolin or Porcelain Clay........ 8
Potters' Clay,........ 8
Crockery Ware made therefrom,........ 8
Caution in using such Ware,........ 8
Baltimore Stoneware,........ 8
Potters' Clay Abundant in Maryland,........ 8
Fire Clay in the Coal Regions,........ 9
Mt. Savage Fire Bricks,........ 9
COAL........ 9
Three Coal Regions in Maryland,........ 9
Potomac Coal Basin,........ 9

Superiority of the Coal,........ 9
Meadow Mountain Coal Basin,........ 9
Youhioghany Coal Basin,........ 10
Caution against wasting money in searching for Coal in any other than the Coal formations shown on the Map (No. 19.)........ 10
IRON ORES........ 10
Bog Iron Ore, on the Eastern Shore,........ 10
Brown Hematite, in Baltimore County,........ 11
" in Anne Arundel County,........ 11
" in Carroll County,........ 11
" in Frederick County,........ 11
" in Washington County,........ 12
SPECULAR IRON ORE AND RED OXIDE OF IRON........ 12
In Frederick County,........ 12
In Alleghany County,........ 12
FOSSIL ORE in do. 12
MAGNETIC IRON ORE,........ 12
In Baltimore, Harford and Carroll Counties,........ 13
CARBONATE OF IRON,........ 13
In Cecil, Harford, Baltimore, Prince George's and Anne Arundel Counties,........ 13
Of the Coal Regions,........ 13, 14
Iron Pyrites or Sulphurate of Iron,........ 14
ORES OF COPPER, LEAD, ZINC, CHROME, MANGANESE, GOLD, &c........ 14, 15, 16
Metalliferous Districts,........ 14
Lead and Zinc in Baltimore County,........ 15
Chrome Ores in Baltimore County,........ 15
First Manufacture of Chrome Yellow,........ 15
Chrome Ores in Cecil and Harford Counties,........ 15
Copper Ores in the same range,........ 15
Copper Mines in Carroll and Baltimore Counties,........ 17, 15
Cobalt Ore in Carroll County,........ 15
Copper Ores in Frederick County,........ 16
Lead Ores in Frederick County,........ 16
Copper Ores in the Catoctin Mountain,........ 16
Copper Ores in Middletown Valley,........ 16
ARTESIAN WELLS,........ 17
Number in this State,........ 17
Districts to which they are Applicable,........ 17
At the Naval School at Annapolis,........ 17
Proposed in the State House Yard,........ 17

NOTE.—In order to avoid delay, I have been compelled to correct the proofs in a hasty manner. The verbal errors which may be found will doubtless be corrected by the reader. Two such errors were noticed whilst making the above table of contents, which the reader will please correct, viz: on page 18, line 11th from the top, strike out "the color of," and in the Appendix, page 4, 11th line from Hydraulic Cement, for "Cumberland." read Chesapeake.

PREFACE.

It is not necessary to discuss the importance of Agriculture, or to indicate the necessity that impels us to seek means for its improvement. The circumstances connected with the progress of civilization and the increase of population, require increasing supplies of food and other products of the soil, for which mankind depends wholly upon agriculture, at least, in civilized countries.

In the early settlement of a country, and whilst the number of inhabitants is limited, a very defective system of agriculture will produce more than a sufficient supply of food. But with a large increase in population, the same imperfect system will no longer suffice. "The land must be better tilled, its qualities and defects be studied, and means gradually adopted for extracting the maximum produce from every portion susceptible of cultivation."

A large portion of Europe is in this latter condition. Within the past eighty years the population of Great Britain has more than doubled, and there has been a corresponding increase of agricultural products. The importation of food from abroad has formed, and continues to supply, a very small fraction of the consumption. There is no increase in the area of land, but by improved systems its products have been more than doubled.

In the older settled States of the Union, the exhausting systems of culture lessened materially the yield of the land, whilst the demand continually increased. And but for the bringing into cultivation the virgin soils of the West, we should scarcely be able in the old States to supply our existing population with food. Even as matters now are, we find that, with the exception of grain and flour, the average value of food, both vegetable and animal, ranges in the cities and in the thickly settled portions of the older States from twenty to one hundred per cent higher than was the case forty years ago.

It is clear, therefore, that production has not kept pace with population in the older States.

Considerations of this kind have impressed themselves upon the minds of the public, and have infused a lively interest in the progress of agriculture with a desire for its improvement. Those more immediately interested are duly sensible of the fact that "knowledge is power," and that to perform any operation to the *best advantage*, requires a knowledge of all the circumstances connected with it. In days long past farmers were, in general, mere laborers; working without thought, (so to speak). But in the progress of time, as their position gradually changed from tenants to land-owners, and as their means for acquiring knowledge increased, they discovered that their occupation required mental as well as physical labor. The natural result has been an increasing desire for further and full information, in reference to those sciences more immediately connected with their pursuits.

An extensive acquaintance with agriculturalists in several of the counties during many years past, has given me abundant evidence of this increasing desire for knowledge. They are fully aware of the importance of knowing *what a soil is*, and also, that this question can only be fully answered by tracing it from its origin and through all its successive changes, both physical and chemical. And further, they are satisfied that this essential investigation can only be properly made, with the aid of geology and chemistry. They also desire this aid in the application of our various resources, suited to the improvement of their soils, and to other purposes.

We have, in fact, abundant evidence of this tendency among the intelligent agriculturalists of our State, to become better acquainted with the sciences connected with their profession: witness the agricultural periodicals and other useful publications in their homes, and their intelligent remarks upon what they read. Another evidence is the establishment of a State Agricultural College. The creation of this institution, so honorable to its founders, proves clearly that the farmers and planters of Maryland are determined that their sons shall not have to toil in "the pursuit of knowledge under difficulties," as has been their own lot. They have, therefore, reared a noble institution for the education of their sons; not only to aid them in cultivating the soil in such manner as to obtain a maximum product, but to give the agriculturalist the rank among men to which he is entitled, but has not yet obtained.

With able and accomplished professors, the College has been opened, and as I had an opportunity to observe during a recent visit, it is in successful operation, under the most favorable auspices. I cannot but believe that so valuable an institution will be fully sustained by our people, and such aid afforded as will increase its sphere of usefulness.

We may confidently expect that when the graduates of the College shall take their positions as farmers and planters, a new era will begin in our State. These young gentlemen will return to their homes with such a training as will make them really practical farmers, knowing well what they are doing, and their light will spread around them.

There is no profession which is intimately connected with so many branches of science as that of the farmer or planter.

In his every day pursuits he is bringing into play the principles of geology, chemistry, botany, entomology, physiology, natural philosophy and mechanics.

Can it be possible that these ought to be ignored by those so deeply interested in their daily application?

The authorities of the State of New York, aware of the importance of the sciences in their industrial applications, caused to be executed a survey of that State, embracing not only its geology, but every branch of its natural history. Nineteen quarto volumes of the final report have been issued, to be followed by at least two more.

It is to be hoped that the day is not distant when Maryland will institute a similar work. My duties are necessarily limited to the application of Chemistry and Geology to Agriculture.

In the present day there are perhaps few who will refuse their assent to the proposition, that the character of the industrial operations of every country, depends for the most part upon its geological constitution, modified of course by climate.

It is equally certain that, in connection with climate, the geological structure and mineral components of any region, determine the character and the fertility of its soils.

The mineral masses which constitute our planet, consists of many different kinds of rocks, clays, sands, and other matters, containing certain substances essential to the growth of plants. Among these are silica, alumina, lime, magnesia, potash, soda, oxide of iron, phosphoric and sulphuric acids, chlorine and other matters.

These substances, or portions of them, exist in much larger proportions in certain kinds of rocks than in others; whilst there are rocks in which matters essential to plants are so firmly united together, and to other substances, that they are very slowly liberated for the wants of vegetation.

At and near the surface of the earth, atmospheric agencies are incessantly disintegrating and decomposing the rocks and other mineral matters. The resulting products remain in part and form an earthy covering. Portions, however, are dissolved and assist in furnishing the mineral matters in spring water, whilst others are washed off during heavy falls of rain and deposited in lower places, or carried into streams and to the ocean.

What we term soil is that portion of this earthy covering,

lying at and near the surface and more or less mixed up with the remains of plants and animals that have died thereon.

There are numerous kinds and varieties of rocks differing from each other both in their chemical constitution and physical structure. Some of them abound in matters required by plants, which in others are either deficient or absent, as will be seen in the description which will be given of the rocks of Maryland.

From what is now known of the origin and characters of soils, we must conclude that the very foundation of an intelligent and practical application of science to agriculture in any region, must consist in a thorough investigation of its geological and mineral constitution.

A survey of this kind for our State should have for its object the determination of the chemical, physical and other characters of each kind of rock, bed of clay, sand, marl, or other mineral deposit within our borders.

They should be minutely described, and their position and extent be accurately shown on a map, and sections upon a large scale. We should make ourselves acquainted with the properties of every mineral that can be usefully applied to the soil, and also, with those that may promote industrial operations within our limits. These last should by no means be overlooked, because of their importance in adding to the demand for the products of the farm.

In addition to many obvious advantages, we shall find if a work of this kind be thoroughly executed, that the wants of our soil can generally be supplied with much less dependence upon manures or fertilizers from abroad than has hitherto been supposed.

Fully impressed with the correctness of these views, I propose to give in this report a sketch of the geological features of Maryland. Much of my time during the last thirty years has been occupied with investigations connected with the industrial application of Chemistry, Geology, &c., in Maryland, Pennsylvania and Virginia. The information obtained whilst thus engaged, is as far as it seemed to be required, embodied in the present report.

Since the commencement of my term of office on the first of May, 1858, I have been fully occupied with the field work portion of my duties whenever the weather and season permitted. Preliminary investigations have been made in nearly all the counties, although not as fully as I designed, because of the very unusual amount of rainy weather during the autumn of 1858, as well as during the last winter and spring. The intervals during which the field work could not be prosecuted in the counties, have been devoted to collecting such information from abroad, as well as at home, as might prove useful to the public. My position also makes it necessary to reply to many inquiries, verbally and by letters to our citizens.

To aid in forming a correct idea of the agricultural and other industrial capabilities of our State, a geological map is an essential requisite, but none such had hitherto been constructed, and I have felt it incumbent on me to commence the construction of one. The first difficulty in the way appeared to be the want of a correct geographical map.

To supply this deficiency as far as possible, I have collected all the useful data within my reach. A large map has been constructed under my supervision by Mr. August Faul, who has executed the work with great fidelity.

I was obligingly furnished by J. H. Alexander, LL.D., with the large manuscript map constructed by him whilst State Topographical Engineer, some twenty years since. This is altogether superior to any published map; but the labors of the U. S. Coast Survey had not then made progress in Maryland, and it was impossible to give the shore lines within the tidal districts correctly, as has since been done under the able direction of Prof. Bache.

I am indebted to the last named gentleman for his large charts of the Chesapeake Bay, and for other charts which enabled us to delineate the shore lines accurately.

There have also been published within a few years, a number of county maps from which, as well as railroad surveys, I have derived much information. To Mr. Martinet I am indebted for the use of his manuscript maps of Howard and Kent counties, which he is about to publish.

Capt. Thos. J. Lee, of the Maryland and Virginia boundary commission, has kindly aided in determining the position on the map of the northeast corner of the State, about which there was much difficulty.

Upon this map it is proposed to place the geological features of our territory, as they can be accurately determined.

The scale of the map is 1 to 200,000, equal to .3168 inch, or rather over 5-16 of an inch to the mile. It is, therefore, of suitable size for placing the final geological results.

An inspection of this map will show that it is infinitely in advance of any existing map of our State. It is, in fact, the best that can be construed out of the materials we possess at this time, whether published or in manuscript. It must, therefore, suffice for our geological purposes until we can have a complete survey of the entire State.

I ask the privilege to suggest the present as a suitable time for authorizing the commencement of a full and complete survey of those parts of the State not covered by the operations of the U. S. Coast Survey. Prof. Bache and his able corps of assistants have nearly completed their work within our limits in the most accurate manner. If Maryland will authorize the survey to be continued over the remaining portions of the State, by means of a triangulation and the requisite plane table work, our citizens can in a few years be supplied with a complete and accurate map of the State.

It has been deemed necessary in order to illustrate the present report in a proper manner, to construct a preliminary map, and I have availed myself of the kindness of Dr. Alexander, in the use of a smaller map, constructed by him some years since, the scale being one-third that of the large map.

Upon this I have delineated the prominent geological features of the State with as much minuteness as is practicable in the present state of the work, and the small scale of the map.

Information is so frequently desired by intelligent farmers, as well verbally as by letter, which it is difficult to give without an aid of this kind to refer to, that I have been encouraged to construct this *preliminary map*, it being the first attempt of the kind for our State.

As I am unwilling to apply the term "Geological map" to any other than one embodying results of a detailed and *final* survey, the title I have applied to it is "Geological Illustrations, &c."

These "Illustrations" will be readily understood with the aid of the "Key" on the right hand side of the sheet.

The formations are numbered, beginning with the older rocks, formerly called "Primary," and following in the inverse order of the relative ages of the formations up to the most recent, which is the post tertiary No. 24. I have added the names applied to the same formations in the final reports of the surveys of New York and Pennsylvania. Each color or shade indicates a distinct formation, with the exception of 14, 15 and 18. Each of these embraces two formations, which run through the State in belts too narrow to be separately shown on a map of this size.

The sections show the relative position of the formations, each of which is believed to be of more recent origin than the one on which it rests.

In order that the "Illustrations" may be published with the report, for the use of the Legislature, I have caused it to be engraved on stone by the Messrs. Hoen, the eminent lithographers of Baltimore, so that it may be printed in colors. To effect this eight separate stones are used, each with one color, but yet by a skilful application of these, one upon the other, we have with eight impressions each of the twenty-four divisions shown by a different shade or color. This is a recent and beautiful improvement in color printing. The Pennsylvania geological map was engraved and printed in Edinburg, Scotland; but the sample here presented shows that we have in our own State those who are able to execute work of this kind with great fidelity and neatness.

It has been suggested that what is called a "population map" should also be presented with the report. The construction is such as will show at a glance the ratio of population to the square mile in each county. Dr. Alexander has kindly arranged the plan for such a map, and the Messrs. Hoen are prepared to furnish the requisite number of copies of both maps upon moderate terms by the time the printing of the report shall be completed.

When the survey of the entire State shall have been made, and the large geological map be completed, it will of course be expected that a final report will be presented for the use of the public.

The final report should be divided into two parts:

1. A full and systematic account of the geology and mineralogy of the State.

2. Economic geology, in which our mineral resources of all kinds shall be described with especial reference to their application to agriculture and other branches of productive industry.

The geological sketch now given is intended to aid in familiarizing our citizens with the principal geological features of their State, and especially to show their connection with its great and leading interest, which I need not say is agriculture.

I have endeavored, as far as possible, to avoid the use of technical terms, and the descriptions are made as popular as is practicable. It would be just as impossible to speak or write about the application of science to agriculture without the use of some of these terms, as it would be for a farmer to describe operations and circumstances connected with his pursuits without the use of the terms usually applied. The farmer may be reminded that these are the technical terms of his profession, and are fully as unintelligible to the uninitiated as are those of chemistry and geology.

As agriculture is now inseparably connected with science, or rather has become a science itself, it is incumbent on those connected with it to understand the few terms which have been adopted from other branches of science.

By reference to a few elementary books, not only would these terms be readily mastered, but much useful information be acquired. Among American publications I would especially recommend the "Elementary Geology" of Professor Hitchcock, of Amherst College, which I consider the best elementary treatise yet published. In reference to Chemistry I may refer to "Wells' Principles and Applications of Chemistry," just published. There is no farmer that would not find advantage in devoting some of his winter evenings to books of this kind. The cost of the first is $1.25, and the second 87½ cents, and they can be had of all the prominent booksellers in Baltimore.

In my opinion the best book upon agricultural chemistry is that of M. J. Pierre, published in Paris. As it is in the French language, is of course not accessible to all, but to those who can read it, I would strongly recommend the book. It is a duodecimo of 523 pages, and not expensive.

CHAPTER I.

Descriptions of the principal minerals of which the rocks of Maryland are composed.

Our earth for the most part consists of rocky masses, many of which "crop out," or become visible at the surface, or are met with in mining or sinking wells.

Although there are many varieties of rocks, yet we find that the number of mineral species considered essential to their composition is quite limited.

In order that the characters of these rocks may be better understood, I shall briefly describe the most important simple minerals which constitute their mass, before entering into a description of the rocks.

1.– QUARTZ.

This substance presents itself under a variety of different aspects, so that the term is more properly applied to a group of what are termed "silicious minerals."

All the varieties of quartz are principally composed of silica, a name which has become familiar to most persons, because of being constantly stated in reports of analyses of soils, manures, ores, ashes of plants, and other matters.

Silica was formerly supposed to be one of the simple elementary substances, but modern chemistry proves it to consist of a peculiar substance called silicium, . 48.05 per ct.
Chemically combined with oxygen, . . 51.95 "

It is believed that silica constitutes more than one-half of the mineral mass of the earth to the depth to which man has penetrated, and it plays an important part in the economy of vegetation. It is one of what are called the inorganic constituents of plants, being always found in their ashes. It will be again referred to in a subsequent chapter.

Silica is considered by chemists to be an acid, because it combines with what are called bases, such as alcalies, earths, and oxides.

Quartz, if perfectly pure, would, in fact, consist wholly of silica, but it has never been met with absolutely pure. The nearest approach to pure silica is in the perfectly transparent crystals of quartz called rock crystal, which in some instances contains 99½ per ct. of silica.

The most abundant variety of quartz, often called "flint," is the vitreous or hyaline kind. The name as applied to the common quartz of the country is obviously improper, because it belongs to the variety of which gun flints are made; and the two differ materially in appearance, structure and origin.

The common or vitreous quartz is the principal stone in our fields in the middle counties of Maryland, and is almost the only constituent of pebbles, gravel and sand. No description is needed of a mineral so well known. It contains from 95 to 99 per ct. of silica, whilst "flint" contains from 86 to 94 per ct.

Another variety of quartz, whose texture and constitution is very similar to that of flint, is called chert or hornstone. It is found in layers and masses in some of the limestones of Washington and Allegany counties.

The variety called pitchstone has a resinous lustre, is much less hard than either of the above, and presents a greater variety of color. It usually contains from 80 to 90 per ct. of silica, 6 to 10 per ct. of water, besides small proportions of oxide of iron and alumina.

There are other varieties of quartz, which as they rarely or never form essential portions of large masses of rock, need not be described at this time. Among these are agate, onyx, chalcedony, cornelian, and amethyst.

Quartz is considered one of the most indestructible substances in nature, and it is for this reason that it is almost the only constituent of gravel and sand.

2.—FELSPAR.

This group of minerals is also widely distributed, and forms a large portion of most rocks of igneous origin, among which are granites, sienites, and the varieties of trap rocks, including hornblende rocks, porphyry, basalt, &c.

The colors of felspar are usually white, inclining to gray, greenish, reddish, or flesh color. It can generally be distinguished from quartz by its laminated structure, and is less hard than quartz. The chemical composition of several varieties will be seen in the subjoined table:

Silica	64.20	67.20	66.73	67.94	61.06	59.60	53.48
Alumina . . .	18.40	20.03	17.36	18.93	19.68	24.28	26.46
Potash . . .	16.95	8.85	8.27	2.41	3.91	1.08	0.22
Soda	.....	5.06	4.10	10.	7.55	6.53	4.10
Lime, oxide of iron, and magnesia. . . .	.45	.....	3.54	.72	7.80	8.51	15.74

Although there is considerable variety of composition, yet we find that potash or soda, or both, are found in all of them.

Felspar becomes an important material in the economy of nature, because it is believed to have furnished the greater portion of the silicates of the alcalies, so essential to the growth of plants, by means which will be adverted to in another chapter.

3.—MICA.

The term isinglass has been often applied to this mineral; improperly, however, because that name belongs to another well known article of animal origin.

Mica is usually gray or greenish, and sometimes black. It may be readily divided into extremely thin plates, by which character it may be distinguished from talc, as the elasticity of a very thin plate of mica is such that if bent it will return to its original shape. This does not take place with talc, which it otherwise resembles.

Mica varies somewhat in its composition, as is shown in the following table of analyses of six varieties:

	1.	2.	3.	4.	5.	6.
Silica	46.23	40.19	50.82	47.50	47.19	36.54
Alumina	14.14	22.79	21.33	37.20	32.80	25.47
Oxide of manganese . .	4.57	2.02	.....	.90	2.58	1.92
Oxide of iron	17.97	19.78	9.08	3.20	1.47	27.06
Potash	4.90	7.49	9.86	9.60	8.35	5.48
Lithicà	4.21	3.06	4.05	.....	.....	.....
Fluoric acid	8.53	3.99	4.81	0.56	.29	2.70
Lime	.....	.....	.....	.....	6.13	0.93
Water	.....	.....	.....	2.67	4.07	.....

It will be noticed that all the above contain potash.

Mica is a constituent part of granite, gneiss, and mica-slate rocks, and also is found in most sandstones and some other rocks, but as it is very slowly decomposed by natural causes, it does not so readly give up its alcalies as felspar.

4.—TALC AND STEATITE.

Pure talc is sometimes white, but usually of some shade of green, and is frequently more or less transparent. It is mostly soft and unctuous to the touch, and has a pearly lustre.

Steatite, usually called soapstone, is a massive and less pure variety of talc.

The composition of talc is as follows:

	1.	2.	3.
Silica	62.58	65.75	64.85
Magnesia	35.40	31.68	28.43
Oxide of iron	1.98	1.70	1.40
Water	.04	4.83	5.22

5.—SERPENTINE.

This like talc consists principally of silica and magnesia, but contain more water than talc, as is shown in the table below:

	1.	2.
Silica	36.19	41.67
Magnesia	21.08	41.25
Oxide of iron	22.73	.....
Water	10.08	13.80
Oxide of chrome alumina	3.06	1.87

Serpentine occurs massive, and its color is usually some shade of green; sometimes variegated, with shades of red, blue, and purple. It is one of the constituents of the much prized verde antique marble.

CHAP. I.

6.—CHLORITE,

Feels soft to the touch, with a lamellated structure, and varies in color from pale to dark olive green. Its usual composition is as follows:

		Chlorite schiste.
Silica	30.01	26.80
Alumina	19.11	19.60
Magnesia	33.15	14.30
Oxide of iron	4.81	23.50
Potash	.....	2.70
Water	12.52	11.40

Chlorite forms the principal part of the rocks named chlorite slates, or talc chlorites, and often contains lime or potash or both.

7.—AUGITE, OR PYROXENE,

Forms a group of minerals consisting of silica, lime, magnesia, and oxide of iron, &c. The color varies from black to green, and sometimes when iron is absent, is nearly white. Its composition varies considerably; though specimens from different localities gave:

	1.	2.
Silica	58.50	54.08
Lime	17.50	23.47
Oxide of iron	4.—	11.02
Magnesia	20.—	11.47

Augite is often a constituent of trap or hornblende rocks, and is found in the dolomites or magnesian limestones in parts of Maryland.

8.—HORNBLENDE,

Structure lamellar, color varying from nearly black to greenish.

The following shows the composition of several varieties:

	1.	2.	3.
Silica	57.60	54.60	53.1
Lime	9 56	10.45	11.4
Magnesia	7.85	19.30	7.4
Oxide of iron	22.67	12.10	25.6
Alumina	.75	85	1.7
Water	.....	1.55	.....

Hornblende is a constituent part of trap hornblende, syenite, and hornblende slate. In some cases it is more or less mixed with granites, gneiss, micaslate, and porphyries.

9.—EPIDOTE.

The color of epidote has usually some shade of a greenish color, varying in fact from bottle-green to greenish gray. Its hardness is about equal to that of felspar, and its structure is usually lamellar.

The chemical composition of few specimens is as follows:

	1.	2.	3.	4.
Silica, ... , ...	37.	39	39.30	40.25
Alumina	27.	26.	29.49	30.25
Lime	14.	15.	22.96	22.50
Oxide of iron	17.	18.5	6.48	4 50
Oxide of manganese	1.5	1.25	.....	.....

10.—CARBONATE OF LIME.

This species presents itself under a variety of forms, such as calcareous spar, and the different varieties of limestones. When pure it consists of—

Lime	56.15
Carbonic acid	43.85

It effervesces in most acids.

The purest form of carbonate of lime is calcareous spar, which is often found in cavities of rocks in regular crystals, as well as in masses, which can be readily cleaved into thin plates, and even into regular rhombic figures. It is often found filling cracks in limestone and other rocks.

11.—DOLOMITE, OR MAGNESIAN CARBONATE OF LIME,

Somewhat resembles carbonate of lime in appearance. They can, however, be distinguished by the fact that the latter effervesces rapidly in acids, whilst the magnesian carbonate effervesces very slowly.

The composition of pure dolomite is as follows:

Lime	30.3
Magnesia, ...	22.4
Carbonic acid, ...	47.3

12.—OXIDE OF IRON.

Iron does not occur in nature in the metallic state except in meteoric stones, but is always combined with other substances.

The protoxide of iron containing 22¼ per ct. of oxygen, and 77¾ of iron does not constitute a distinct mineral, but is always in chemical union with silica or other acid. Thus, in mica, chlorite, hornblende, epidote and other minerals, we find silica combined with alumina, lime, magnesia, alcalies and oxides of iron, and small proportions of manganese. The iron in most minerals and rocks is in the state of protoxide, but it has a strong affinity for a larger proportion of oxygen. The absorption of this from air and water, alters the character of the minerals and assists in disintegrating rocks.

The magnetic oxide of iron contains 72.42 of metal, and is an important ore of iron. It is found in large masses in veins and seams in rocks; and sometimes disseminated in the form of crystals and grains in hornblende and augitic rocks, in steatite and in chlorite.

The peroxide to which the name of sesquioxide of iron has been applied, contains 70 per cent. of metal and exists in large masses, beds, seams and in veins. It also occurs disseminated in some of the shales and slates. The enormous masses in which it occurs, in the island of Elba and in Missouri, have induced some writers to class it among the rocks.

The color of some varieties is red, but it has usually a brilliant metallic lustre and a dark grey color. Its powder is always a dark red, which distinguishes it from the magnetic oxide; whose powder is black.

Another form of oxide, the peroxide (chemically united

with about 14 per cent. of water,) is abundant, although it does not form an essential constituent of rocks. It constitutes the rust of iron and is formed when the silicates and other salts of iron become altered by exposure and disintegrate.

It constitutes the ore called brown hematite, which forms extensive beds and masses. It is common in the upper portions of metalliferous veins under the name of gossan.

13.—OXIDE OF MANGANESE,

Exists as a silicate in numerous minerals and rocks, and also in masses and in veins.

Both iron and manganese are so extensively distributed over the surface of the earth, that there are few rocks or clays in which they do not occur, either as essential or adventitious ingredients.

14.—SULPHUR,

Occurs in volcanic regions, and in combination with metals is extensively distributed. It is most abundant as sulphuret of iron or pyrites which occurs in veins and in masses, as well as disseminated in various rocks, whose destruction it hastens when exposed to atmospheric action. The oxygen of the air converts it into acid sulphate of iron, or copperas, which flowing over other minerals affects chemical changes therein. When it comes in contact with limestone or marl it produces gypsum or sulphate of lime. It is by this means nature often supplies gypsum to the soil.

Sulphur or its sulphurets or sulphates, are so universally distributed that there is no soil where plants can grow, in which at least a trace of one of them cannot be detected.

15.—PHOSPHORIC ACID,

In some of its combinations is also in every soil, because there can be neither animal life nor vegetation without its aid. Phosphates of lime, alumina and iron are found among minerals. Phosphate of lime, although more abundant than other phosphates, occurs in few localities in large quantities as a mineral. We find it in a few mineral veins, and in small proportion in certain marls derived from fossils. The beds of phosphate of lime in the more recent formations are evidently of animal origin as in the case of the coprolites of Europe.

16.—Chlorine.

This substance combined with soda, forms common salt, which exists in immense beds or strata in the new red sandstone formations of Europe and Asia. It has also been found in the silurian formations of the southwestern part of Virginia. The waters obtained by boring into the rocks below the coal formation in New York, Pennsylvania, Virginia, &c., abound in common salt. With these exceptions, although universally distributed, it occurs in minute proportions in most rocks and in all spring water, even in the purest.

CHAPTER II.

Mineral Characters of Rocks.

For convenience in referring, these rocks may be classified as follows:

A.—Rocks generally considered of igneous origin.

1. Granite.
2. Syenite.
3. Massive Quartzite.
4. Porphyry.
5. Amygdaloid.
6. Trap, including Hornblende rock, or Amphibolite.
7. Serpentine.

B.—Rocks of aqueous origin.

(a) Chemical deposits.

1. Limestone.
2. Dolomite or Magnesian limestone.

(b) Mechanical or sedimentary deposits.

1. Sandstone.
2. Conglomerate or Puddingstone.
3. Breccia.
4. Clay slate.
5. Shale.
6. Clays.

(c) Metamorphic rocks.

1. Gneiss.
2. Mica slate.
3. Hornblende slate.
4. Talc slate.
5. Chlorite slate.
6. Quartzite.

7. Limestone (granular).
8. Dolomite.

A.—Rocks of Igneous Origin.

The rocks of this class give no evidence of stratification.

1.—Granite.

This was considered by the older geologists to be the oldest of the whole series of rocks. It is composed of quartz, felspar and mica, forming what is called an aggregate. By this it is understood that each of its constituents exists in separate crystalline grains firmly cemented or aggregated together, forming a solid mass.

We have seen that both felspar and mica, and even quartz, vary considerably in chemical composition and in color. We would expect, therefore, a great variety of appearance in granite.

In some localities the grains are extremely small, so as to be scarcely distinguished by the most practised observer, without the aid of a lens. Again we have granites whose components are several inches in diameter; and there is every grade between these two extremes.

The most common color of granite is some shade of grey given to it by the dark color of the mica, and sometimes also of the quartz. It sometimes contains red felspar. When all the materials are white, or nearly so, as in some localities, the granite is almost white and strongly resembles granular limestone. In one instance, which came to my knowledge, a granite of this kind was actually tried in a limekiln under the supposition that it was limestone! A few drops of acid, or even trying its hardness by scratching it with the point of a knife would have shown the blunder.

Granite abounds in Baltimore county, and also occurs in Cecil, Harford, Carroll, Howard and Montgomery.

The finer grained varieties can be readily dressed with the hammer and chisel, and it can be split off into stones of any required dimensions. It is much used in constructions requiring strength and solidity.

Indestructible as granite seems to be, yet it is disintegrated under certain circumstances, or becomes rotten in common parlance. This is principally owing to the action of water containing carbonic acid, which dissolves out the silicate of potash contained in the felspar, the texture of which is thus destroyed.

This is one means by which the silicates of potash, soda and lime, so necessary to plants, are supplied to the soil.

When, as is sometimes the case, granite is almost wholly composed of felspar, by the removal of its silicate of potash, the remainder of the mass is converted into kaolin or porcelain clay, which is used in the manufacture of true porcelain.

There are known localities of this in Harford, Cecil and Baltimore counties, and doubtless it can be found in other counties in the same geological range, if they be carefully explored.

2.—Syenite.

This rock is also an aggregate, whose essential components are felspar and hornblende. It usually also contains grains of quartz. It has a speckled appearance, owing to the dark colors of the crystalline grains of hornblende being intermixed with those of the felspar. When the felspar has a reddish tint and is in large proportion, the rock appears to have a red color.

In some localities it seems as it were to pass into granite by the disappearance of the hornblende, whose place is taken by mica and a larger proportion of quartz. On the other hand, from a diminution of the proportion of felspar and an increase of hornblende, it passes into amphibolite or hornblende rock. Syenite is subject to disintegration like granite, and has similar architectural applications.

3.—Quartz.

Although quartz sometimes occurs in sufficiently large masses to come under the definition already given of a rock, it never forms such extensive areas as granite and syenite. In some parts of this State it constitutes dykes, veins or masses, in metamorphic rocks, especially in mica slate.

Its characters has already been noticed under the head of simple minerals.

4.—Porphyry.

There are many varieties of Porphyry, all differing from granite materially in structure.

A porphyry is made up of crystals (generally of felspar,) which are imbedded in what seems to have been a soft paste, but subsequently hardened. This *paste* is said to constitute the base of the porphyry, and is usually compact so as to present a smooth surface when fractured. It is mostly a variety of compact felspar differing somewhat from that constituting the imbedded crystals and less pure.

The colors both of the base and the crystals in different localities are various.

This rock is by no means abundant in Maryland, being only found in the Catoctin Mountain, and in smaller quantity where the Baltimore and Washington Railroad crosses the Patapsco river. There is also a thin dyke of porphyry near the Relay-house, on the Northern Central Railroad.

5.—Amygdaloid.

The base of this rock is often similar to that of porphyry, but the imbedded material, instead of being in crystals, consists of rounded pieces similar in shape to the kernel of an almond. Hence the name.

The imbedded matters are usually carbonate of lime, or chalcedony, supposed to have been deposited in pre-existing cavities in the rock.

This rock is rare in Maryland, having only been found in the Catoctin Mountain.

6.—Trap Rock. (Amphibolite.)

This term is by some geologists applied to various intrusive rocks, but we shall restrict it for the present to those in which hornblende is an essential constituent. It consists of hornblende and felspar, and sometimes contains black or brown pyroxene.

These minerals are usually in crystals aggregated together. In some cases the crystals are large, where the rock resembles, and in fact seems to pass into syenite, as near the base of the Catoctin Mountain, northeast of Emmitsburg. In other localities amphibolite consists almost entirely of hornblende. Sometimes, as in several parts of Frederick and Carroll counties, the crystalline grains are so small as to give the rock a compact appearance, and the aggregated constitution of the mass can only be determined by a good lens.

Different varieties of amphibolite or trap exist in Cecil, Harford, Baltimore, Carroll, Howard, Montgomery, and to a small extent in the eastern part of Washington county.

Some varieties of porphyry, like the granites and syenite, readily disintegrate, and form soils of medium fertility, whilst others are acted on so slowly as to afford only a thin covering of soil.

7.—Serpentine,

Has already been noticed as a simple mineral, but must also be enumerated among the rocks. It seems to have been forced up as an intrusive rock in isolated masses of limited areas in Cecil, Harford, Baltimore, Howard, and Montgomery counties. It is slowly acted on by atmospheric agents, and furnishes a barren soil wherever it is the nearest rock to the surface.

B.—*Rocks of aqueous origin.* A.—*Chemical deposits.*

1.—Limestone.

The term limestone is applied to a very important class of rocks, differing much in appearance from each other. They are essentially composed of carbonic acid and lime, mixed up with various other mineral matters, which may be termed impurities.

Limestones are supposed to have been deposited from aqueous solutions, and when unaltered by heat, are usually more or less compact and fine-grained in appearance. They have also usually a foliated or slaty structure.

The action of internal heat has in many cases rendered them more compact, and often obliterated the foliated structure in a greater or less degree. Again, we have limestones made up of aggregates of chrystalline grains of carbonate of lime and small portions of other minerals disseminated through the mass. Some of these are called saccharoidal, because of their resemblance to loaf sugar. Those consisting of large crystalline grains have the local name of alum limestone. All these varieties are abundant in Maryland.

2.—Dolomite, or Magnesian Limestone.

Pure dolomite has been noticed among the simple minerals. The granular variety is rarely found in quantity as a rock, without being mixed with various impurities. It occurs in this state associated with metamorphic limestones in Harford, Baltimore, and Howard counties. In addition to its occurrence in this way in large masses, we have the carbonate of magnesia more or less mixed up with all our limestones, especially eastward of the North Mountain. The proportion of magnesia in these mixed limestones varies from one ten or fifteen per cent.

B.—*Mechanical or sedimentary deposits.*

1.—Sandstone.

The essential components of sandstone consist in grains of sand composed of quartz, or other hard minerals, which have been deposited from water, and afterwards cemented together so as to form solid masses.

Sandstones occur in some localities composed almost entirely of nearly pure and white grains of quartz sand, with pure siliceous cement. Such are the white sandstones of which there are several formations in our State, which will be noticed presently.

More frequently, however, there are divers other minerals, including oxides of metals disseminated through the stone, which produce the many varieties of color in sandstones. When sandstones are very fine-grained, and more or less mixed up with fine earthly matter, they are called slaty sandstones.

We have them of every shade of gray, slate color, brown, yellow, and red, usually more or less dull. In some varieties of the slate color, gray and brown, the color is in part owing to carbonaceous matters. The color of yellow sandstones is mostly due to the presence of hydrous peroxide of iron, whilst the red is colored by anhydrous peroxide of iron.

There are also green sandstones whose color is given by silicate of protoxide of iron. These become yellowish and brown by weathering.

Oxides of iron often form a large proportion of the cement of sandstones, and they are rarely free from the oxides of manganese.

2.—Conglomerate or Puddingstone.

When instead of sand the rock is mainly composed of pebbles, whose interstices are filled with grains of sand, and the whole are cemented together, the above names are applied.

These pebbles are mostly quartz, but we sometimes find them to consist in part of granite, gneiss, amphibolite, and other rocks and minerals, which have resisted attrition.

In some localities there are also conglomerates, consisting of the remains of limestones. We have in Frederick county one formation of that kind, which is the better known from the fact of the use of the stone in the old Representative Hall at Washington.

It consists of fragments of limestone varying from the size of a pea to that of a man's head, with here and there one of hard red sandstone, the whole held together by means of a calcareous and feruginous cement. Its component fragments have been but slightly rounded by attrition, so that it approximates in character to that of a breccia.

3.—Breccia.

This consists of parts of other rocks, which by some natural causes have been broken into small fragments, their arrangement being disturbed and afterwards cemented together without being rounded by rolling against each other in swift running water.

It abounds in Italy and in some other countries, but does not exist in Maryland, unless the name shall be eventually applied to the calcareous rock above noticed.

4.—Clay Slates, (Argillite.)

This rock is supposed to have been formed by the deposition of fine mud, or sediment, from water, and consists of the debris of other formations, from which such water flowed. We would expect, therefore, a considerable variety in composition.

They appear, however, to consist principally of mixtures of silica chemically united to alumina and other earths.

Three European varieties, which have been analyzed, give the following results:

	1.	2.	3.
Silica	48.60	60.40	71.
Alumina	23.50	21.40	15.30
Oxide of iron	11.30	6.20	9.30
Potash	4.70	4.60	.....
Magnesia	1.60	.20	.....
Lime	.....	.20	.....
Water	7.60	7.00	3.30

We observe that the first and second are rich in potash, which is absent in the third. This fact materially affects the character of soils, which result from their disintegration. The third cannot but produce a barren soil.

There are also calcareous slates, some of which are rich in lime.

Clay slates are found of various colors, owing principally to the proportions and state of oxidation of the metals they contain. Those of a slate and of a lead color are found usually to contain carbonaceous matter of vegetable origin.

5.—Shales.

The origin of Shales is similar to that of slate, and they are generally of similar chemical composition. They mainly differ from the fact, that although both are often solid rocks at some distance beneath the surface, yet the slates are more apt to retain their form when exposed to sun, rain and frost, &c., while Shales are readily disintegrated when exposed to these atmospheric agencies. Perhaps, also, as a general rule, Shales contain more lime than slates, which would hasten their crumbling.

There are some Shales which alter so rapidly as to be converted into clay in a few years. These varieties are especially among the shales of coal regions, which often contain much bituminous or carbonaceous matter.

6.—Clays.

The origin of these is similar to slates and shales, and they contain similar mixtures of silicates modified, of course, by the geological composition of the region from which the sediment has been derived.

Although some varieties of clay appear to be quite solid and firm under ground, yet they all can be mixed up with water and become more or less plastic. In this they differ from slates and shales, which do not become mixable with water unless they be previously ground to a powder.

The clays of Europe, as well as such as have been carefully examined in this country, are in almost every instance found to contain potash, and it has been shewn also that they have the property of absorbing ammonia and retaining it so firmly that it cannot be dissolved out by water.

C.—METAMORPHIC ROCKS.

This term is applied to those geological formations which are supposed to have had an aqueous origin, and subsequently modified in a greater or less degree by heat and other causes.

1.—Gneiss.

This rock, which is largely developed in the central portions of this State, is essentially composed of quartz, felspar and mica. It differs from granite in containing usually more quartz and mica and less felspar, and in being a stratified rock; whilst granite is ranked with the igneous intrusive rocks which appear to have been forced up from great depths in a liquid form.

Although there is usually ample evidence of stratification in Gneiss, yet in some localities it has been so much altered by the joint action of heat and the intrusive forces from below, as nearly to have obliterated its stratification planes, so as to resemble granite. In certain localities the strata appear to have been bent and folded up in a very confused manner.

Examples of well developed gneiss regularly stratified may be seen in the quarries on Jones' Falls, near Baltimore, and in other parts of the same geological range.

The fissile structure of gneiss is owing to the small plates of mica being arranged parallel to each other, and when these are in large proportion, the rock has more and more of a slaty structure, and in proportion as the felspar and quartz lessen in quantity, it approaches in character to mica slate. It is sometimes, in fact, not an easy matter to determine which of the two rocks a specimen belongs to, as they seem to pass into each other.

On the other hand we have localities in which hornblende replaces, to a greater or less extent, the quartz and mica, and the rock appears to pass into hornblende slate.

2.—Mica Slate.

This rock is essentially composed of grains of quartz and mica, and has always a more or less distinct slaty structure. When the proportion of mica is small, it sometimes forms a hard, durable stone. Varieties containing a very large proportion of mica are often called micashiste, which readily crumbles down.

Some of its beds which contain hornblende and felspar, pass into gneiss.

As we approach the northwestern borders of this formation, in Maryland, (see the map,) the quartz lessens in quantity and the spangles become extremely small. The rock, in fact, passes by insensible shades of difference into talcose slates.

3.—Hornblende Slate,

Appears to be made up of flattened crystals of hornblende, with felspar and frequently quartz. It varies in color from nearly black to dull green, and has always a distinct foliated structure. Its geological position is among the mica slates and gneiss.

4.—Talcose Slate.

I am not willing, at present, to propose a new name for this rock, although there are good reasons to be dissatisfied with that above used, and which has been long in use.

The name seems generally applied to a formation, intermediate between mica slate and argillite, and is the highest or nearest in the series of what are called metamorphic rocks, unless we place argillites in that division.

Talcose Slate appears to have been an argillite which has been altered by heat or other means, so as to assume a more or less glistening appearance. In some cases it has what is called a satin lustre, owing to the presence of scales of mica, or talc, or both, which are so minute as not to be distinguished but with the aid of a microscope.

5.—Chlorite Slate.

This rock is essentially composed of chlorite in flattened crystalline grains and of quartz. Its color is some shade of olive green.

In Maryland it occupies a subordinate position within the mica slates.

6.—Quartzite.

This term is applied to all formations of quartz among the intrusive or metamorphic rocks, when the quartz is in sufficient mass to be considered a rock. It is doubtful whether we have such in Maryland. There are, however, thick veins and masses of quartz in some of our mica slates, to be noticed in another place.

7.—Granular Limestone.

This rock is usually found in contact either with intrusive or metamorphic rocks.

It is composed of crystalline grains of carbonate of lime aggregated into a solid rock, sometimes nearly pure, but most frequently mixed with more or less magnesia and other impurities.

This rock furnishes white marble, including the fine grain statuary marble from Tuscany.

In some localities the crystalline grains attain the size of half an inch in diameter, and such is called, in Maryland, alum limestone.

8.—Dolomite, or Magnesian Limestone.

This term is applied to limestone whose composition is that given to the simple mineral of that name already noticed. The dolomites of Maryland are usually in smaller grains than granular limestone and have a more glistering lustre.

There are often small scales of talc or mica disseminated through the rock, and white augite sometimes is also present, with other minerals.

In addition to the essential constituents of rocks, later chemical researches shew that many ot them contain minute traces of phosphoric acid and chlorine. They have been detected in limestones, hornblende rocks and granite, as well as in slates and shales and other rocks.

CHAPTER III.

GEOLOGICAL FORMATIONS IN MARYLAND.

A prominent feature in our geology is that most of the formations exist in nearly parallel ranges whose course is about N. E. by N.; passing into Pennsylvania on the north and into Virginia on the south and west.

In the completed reports of the surveys of New York and Pennsylvania, different names have been adopted, but I do not feel disposed, at present, to commit myself fully by the adoption of either nomenclature.

The final report upon the Geology of Pennsylvania, with the large geological map, published eighteen months since, is highly creditable to that State, and also to Prof. H. D. Rogers, formerly of Harvard University, but now a Professor in the University of Glasgow.

The work was performed by that gentleman with an able corps of assistants.

Unwilling to accept the local names, adopted for what is called "The New York Geological System," or those of Europe, Prof. Rogers invented a new set of names for those in the subjoined table, numbered from 8 to 21 inclusive.

For the present, perhaps, it will be better to number our formations provisionally. But in order to facilitate comparisons with other systems, I add, in the annexed table, a column for that of Pennsylvania, one for the New York system, and a third for the nearest equivalents in Europe.

The names of those numbered from 1 to 6, inclusive, as well as No. 21, 23 and 24 are common, both to this country and Europe, and have therefore no synonyms. Formation No. 22 is believed to exist nowhere but in our State, and must in due time have a Maryland name.

Most of them, as is shown in the map, constitute narrow belts, and the sections will assist in showing why this is the case. They were doubtless deposited, one upon the other, in succession, and afterwards upheaved by forces from below, which appear to have acted, along lines, nearly coincident with their range or *strike* as it is usually termed.

These upheaving forces seem to have acted most frequently and with the greatest energy upon the metamorphic rocks, 5 and 6, and with somewhat less energy upon the formations numbered 8 to 12, which extend westward to the North Mountain. When we examine the stratification of the rocks from 13 and upwards, we have evidence of less disturbance as we proceed westward. Since the deposit of the coal formations westward of Dan's Mountain, the upheaval seems to have taken place with more uniformed force over large areas. The result is, therefore, that whilst elevations of more than 3000 feet have taken place in some of the ridges, the strata are not generally turned up at such high angles, nor are they so much plicated or broken as in the older rocks.

In describing the formations of our State, we shall commence with No. 5, and then notice the intrusive rocks Nos. 1, 2, 3, and 4, which occur within the limits of No. 5.

FORMATION No. 5.

Gneiss, Mica-slate, and Hornblende-slate, including the Intrusive Rocks 1, 2, 3, *and* 4, *and a portion of the Limestone No.* 11.

For the mineral character of these rocks, reference may be made to the description in Chap. II.

The southwestern limits of these rocks above the tide level, are at the head of tide-water upon most of the streams crossed by the old post road from Elkton via Havre-de-Grace and Baltimore to Washington. Along this line it passes under the cretaceous clays 21 and 22.

It constitutes a belt varying in width from 12 to 20 miles, and extending through portions of Cecil, Harford, Baltimore, Howard, and Montgomery counties, and is bounded on the northwest by Talcose slates, (6,) or more correctly speaking it seems to pass into that formation by insensible shades of difference.

It would be proper to describe the positions of each of the three rocks now included under No. 5 separately, because of the marked difference of soil they produce, but this cannot be attempted until a minute geological survey of them shall have been completed. In the present state of our knowledge they may be viewed as a somewhat confused assemblage, which it will require time and patience to unravel.

Nearest the southwestern limits the prevailing rock is gneiss, with occasional intercalations of mica-slate, and still more of hornblende-slate. The proportion of gneiss is greatest in Baltimore county.

Mica-slate increases in quantity as we proceed northwest, and in Montgomery county it is a prevailing rock. The intercalations of hornblende-slate are more abundant in the gneiss than in the mica-slate.

The accidental minerals in gneiss are few in number in this State, and insufficient in quantity to affect materially the character of the soils produced from this rock. There are numerous minerals, however, of much scientific interest that should be described in a final report. Those of industrial importance will be noticed in the chapter upon the mineral resources of Maryland.

Most of the localities of intrusive rocks are shewn as far as is practicable in a map on so small a scale; in order to illustrate the geological character of this range it is proposed to locate on the large map the *exact* boundaries of these and of the numerous masses too small to be noticed in the "illustrations."

The metamorphic limestones No. 11, which exist in this range, are confined to Harford, Baltimore, and Howard, and include several minor ranges of dolomite. They are associated with a certain kind of mica-slate, which is continuous into Montgomery, and although so far supposed to be unaccompanied by the limestone, yet the indications are so favorable to its existence in that county that it is my intention to make a most minute examination, specially for the purpose of determining the point.

Near the northwestern limits of the mica-slate there are many intercalations of chlorite slate, which is included within what we may term a metalliferous range, in which there are both iron and copper ores, besides cobalt and gold to a small extent.

In some portions of the mica-slate we find garnets, staurotide, and cyanite disseminated in considerable quantity, and as these minerals are very slowly acted upon by atmospheric agents, the rock decays very slowly, and when they form a large proportion of the mass, furnish a light barren soil.

FORMATION Nos. 6 and 7.

Talcose-Slate, including Roofing-Slate.

Allusion was made in the last chapter to the gradual passage of mica-slate into talcose-slate.

In point of fact, we find where these formations approach each other, that the spangles of mica diminish in size, and lose their distinctive characters by degrees, so that there are localities where it is impossible to determine to which of the two the rock belongs. The fine satinlike lustre of the talc-slate becomes more and more apparent, until at length the characters of the rock become clearly apparent.

In studying these talc-slates from their junction with the mica-slates, we find their talcose type becomes less apparent as we cross them in a northwest course, and the fertility of the soil seem greater also.

This formation occupies a small area in the northwest part of Harford, and thence ranges through Baltimore, Carroll, Howard, and Montgomery counties in a wide belt.

With the exception of the southeastern edge, (which may be included within the metalliferous range or the borders of No. 5,) this formation is singularly destitute of minerals other than the rock itself, until we cross the summit at Parr's ridge. In fact the characters of this formation westward of this summit differs so much from that on the eastern side, that it is quite probable further explorations will shew the propriety of making a division in the classification.

The metamorphic limestones between Parr's ridge and the western limit of No. 6, differ materially from those in No. 5. The latter, with the exception of the "alum limestone," (which occupies small areas in Baltimore county,) is generally much mixed up with small grains of quartz, mica, talc, and other minerals. The former, however, is remarkable for its purity, and has a fine grain, and in some of the quarries there are ledges much resembling statuary marble.

It seems, however, that those in the same range, between Liberty and the railroad near New Market, have been more fully metamorphosed and are less pure.

These limestones do not constitute continuous belts, but interrupted ledges, varying in length from a few hundred yards to several miles, and constitute an important element in the agriculture of that region.

The rocks so far enumerated contain no remains of animal and vegetable life. Those which follow contain such remains, and are therefore supposed to have been formed since organic life began on our globe.

FORMATION No. 8.

(Primal of Pennsylvania survey.) Pottsdam sandstone of New York.

This division includes—

1. A hard sandstone made up of grains of quartz, with occasionally grains of felspar and kaolin. The silicious cement seems to have completely filled up the interstices between the grains, so as to give a firm compact structure to the rock. Portions of this rock seem to have been subjected to such changes as to render it doubtful whether it should not be considered a granular quartz, and be classed among the metamorpic rocks. Vegetable life seems to have commenced at the period of the formation of this rock, because it contains fossilized stems of plants.

2. A slate varying in color from gray to brownish and greenish. It is ranked as an argillite, but portions of it assume a marked talcose appearance, especially in the Catoctin Mountain, and in parts of Middletown valley, where it has been much disturbed and altered by proximity to intrusive

rocks. These last consist of amphibolites, (trap,) porphyries, amygdaloyd, serpentine, and epidote. This last named rock is extensively developed both in large masses and intercalated between the slates, and has largely contributed to produce the highly fertile soil of Middletown valley.

Approximate measurements of the thickness of these strata have been made in New York and Pennsylvania, but I have not as yet been able to obtain reliable information upon this point in our State.

FORMATION No. 10.

Auroral series of the Pennsylvania reports. Chazy and Black River Limestones of New York.

This constitutes the most extensive limestone formation of the United States, or in the world, its entire length being over 600 miles. A small area of it occurs in Frederick county, near the Monocacy river, reaching from the Potomac to a point north of Woodsboro', where it is covered by the mezozoic or new red sandstone. Between the South and North Mountains, in the Hagerstown valley, it is developed to the extent of more than three-fourths the area of the part of that fine valley within our limits.

It varies in color from blueish black to blue and gray, etc., and has a compact structure, sometimes inclining to slaty. Some of its layers are nearly pure, whilst others contain from 10 to 30 per cent. of magnesia. The slaty varieties contain variable proportions of other earthy matters.

The lower beds of the limestone have been named calciferous send rock in New York, and calcareous sandstone in Pennsylvania, but I have seen none yet in this State to which these names can be properly applied.

Near its eastern limits in Washington county it assumes a coarse slaty appearance, and some of the layers contain a large proportion of oxide of iron. This is probably the equivalent of the calcareous sandstone of Pennsylvania.

The soil resting upon this formation is among the best in the State.

Fossils are rare, and embrace what are probably among the remains of the first created animals of North America.

FORMATION No. 11.

Metamorphic Limestone.

There are two ranges of these in Maryland, both of which have already been noticed:

1st. In the gneiss and mica-slates.

2d. In the western parts of the talcose-slates of Carroll and Frederick counties.

FORMATION No. 12.

Matinal series in the Pennsylvania reports, Hudson river Slates, Utica Slate and Trenton Limestone in the New York reports.

Not having had sufficient opportunity to investigate this formation in our State, I avail myself of the description given by Professer Rogers for the adjacent parts of Pennsylvania, which is as follows:

1. "A dark blue and blueish gray, soft argillaceous limestone, alternating near its upper limit, with blue calcareous shale." It contains many fossil shells.

2. "A blackish and dark blue fissile slate, usually very carbonaceous, and containing fossils."

3. "Blueish gray shales and sandy slates, containing in their upper portion especially, many beds of argillaceous sandstone, and some layers of a dark gray siliceous conglomerate." It has many fossils.

This formation occurs in Maryland only in the western side of the Hagerstown valley, and although the soil it produces is naturally less fertile than the adjacent limestone, it can be readily made productive if judiciously treated.

FORMATION No. 13.

Levant series of the Pennsylvania report, Oneida Conglomerate and Medina Sandstone of the New York reports.

This formation is divided by Rogers as follows:

1. A compact greenish gray massive sandstone.
2. A soft red sandstone and shale.
3. A hard white and light gray sandstone in thick massive beds, alternating in its upper parts with beds of greenish shales similar to the next formation above, (12.)

In the Pennsylvania report Professor Rogers includes the rocks of this division under the name of the Levant series. In the New York reports the gray sandstone is called Oneida conglomerate, whilst the red and the white sandstone beneath is named Medina sandstone.

These rocks form the summits of the North Mountain, Tonaloway Hill, and Will's Mountain, and from thence dip under Dan's Mountain, and do not again rise to the surface in Maryland.

CHAP. III.

FORMATION No. 14*a*.

Surgent series of the Pennsylvania reports. Clinton group of the New York reports.

The importance of this group is owing to its valuable beds of iron ore, which has caused it to be fully explored in several of the States which it traverses. It consists of—

1. A series of olive, brown, and yellowish slates, with some sandy layers. Some of them assume a claret color by exposure. It contains fossil stems.
2. Iron sandstone, alternating with a greenish sandy shale. Some of the layers contain sufficient iron to be worked. They are of a dark reddish-brown color.
3. Upper shale of a greenish color which changes by exposure to buff colored, and sometimes brown and brownish red.
4. Lower ore shale, greenish, with layers of limestone.
5. A gray calcareous sandstone in thin layers.
6. The upper ore shale consists of blueish and greenish shales, with alternations of thin beds of both pure and shaly limestone, and some beds of calcareous sandstone.

The lower part of this shale contains on both flanks of Will's Mountain, two, and sometimes three, beds of what is called the fossiliferous iron ore, which continues through both Pennsylvania and Virginia, and is largely used in the furnaces in the former State.

7. A thick mass of red shales sometimes slightly calcareous.

A reference to the section will show that this formation, resting upon the sandstone, (No. 13,) appears on the western flank of the North Mountain, and in the principal valleys between that and Martin's Mountain. From thence it is covered by other and more recent formations, until we reach the base of Will's Mountain, where it crops out, and renders the fossiliferous iron ore accessible. Reappearing on the west flank of Will's Mountain, it dips under Dan's Mountain, and does again come up to the surface in this State west of that point.

FORMATION No. 14*b*.

Scalent series of the Pennsylvania report. Onondago Salt Group and the Water-line Group of the New York reports.

It consists—1st. Of blueish, greenish, and red calcareous shales, with some beds of limestone, resting upon a thick series of gray, greenish, and blueish calcareous shales, with beds of impure limestone, and seems to graduate into the cement rock above.

2d. A blue limestone, with bands of chert. It is usually in moderately thin layers, and contains a little magnesia.

FORMATION No. 15*a*.

Pre-Meridian Limestone of the Pennsylvania report. Lower Helderberg Limestone, New York do.

This limestone varies in color from gray to blue, and also varies considerably in composition. Some of the layers are very pure limestone, whilst others are shaly and siliceous. It abounds in layers of chert, which near the western base of the North Mountain are of considerable thickness. It contains numerous remains of shells as well as of corals.

It appears resting upon No. 15*b*, near the western base of the North Mountain, and dips under No. 16. It crops out on both flanks of Tonaloway Hill, and passes under Sideling Hill, Town Hill, Ragged Mountain, Warrior Mountain, and Evit's Mountain, and crops out in the city of Cumberland. We again find it on the west side of Wills Mountain, and dipping under Dan's Mountain again rises to the surface in the State of Ohio; its identity being fully established by its fossils.

These rocks rest immediately upon the limestone of No. 14*b*, and it is possible that we may hereafter determine that both belong to the same formation.

Their range in Maryland may be inferred from that of 14*b*, already inferred.

FORMATION No. 15*b*.

Meridian series of the Pennsylvania report. Oriscany sandstone of the New York report.

A dark colored slate forms the lower beds of this division in some parts of Pennsylvania, but I have not yet met with it in Maryland.

The sandstone is usually of a yellowish color, has a coarse open texture, is more or less calcareous, and contains numerous fossils. It rests upon the limestone last described, (15*a*.)

FORMATION, No. 16*a*.

Cadent Slates and Shales of the Pennsylvania reports. Marcellus Slates, Hamilton group and Gennessee Slates of New York reports.

This group consists of—

1. A black bituminous Shale.

2. Olive, brownish and gray Shales, in some localities containing thin beds of brown and gray sandstone.

3. A brownish and blueish Slate.

Among its fossils are remains of plants allied to those of the coal formation, and the lowest yet known of that kind.

FORMATION No. 16*b*.

Vergent series in the Pennsylvania reports. Portage flags and Ithica and Chemung groups of New York reports.

It consists of—

1. A fine grained sandstone (more or less argillaceous,) in thin layers, parted by thin seams of shale, with marine fossils.

2. Blue gray and olive colored shales with occasional layers of brown and gray sandstones.

The rocks of this group, with those of 16*a*, are largely developed between Licking creek and Hancock, in Sideling Hill, and in all the ridges westward to Cumberland, where they crop out. They are again seen in Dan's Mountain, near its base, dipping westward, and do not appear again in Maryland.

FORMATION No. 17.

Ponent series of the Pennsylvania reports. Catskill group of the New York reports.

This group is the equivalent of the old red sandstone of Europe, and which has been popularized, so to speak, by Hugh Miller. The name of old red shales would be more proper in our State, because it consists principally of rather soft red shales, with a very few beds of red, brown and gray sandstones. Like all the red shales in this State, it contains very few fossils, and these are in a few calcareous layers. It occurs in Sideling Hill and Town Hill, and again on the east flank of Dan's Mountain, under which it dips and re-appears on the west flank of Savage Mountain. It occupies a large portion of Allegany county, west of Savage Mountain, between the three coal basins, as will be seen by referring to the map.

FORMATION No. 18*a*.

Vespertine series of the Pennsylvania Survey. (It does not occur in New York.)

This group consists of greenish and olive colored and blueish shales, with a few thick layers of dull gray sandstones. It contains remains of plants approximating those of the coal formation.

FORMATION NO. 18*b*.

Umbral series of the Pennsylvania reports. (It does not occur in New York.)

It consists principally of red shales with some of an olive green color. There are also occasional alternations of slaty sandstones.

It contains a bed of limestone from forty to fifty feet thick, which is important from the fact that it is the only thick bed of limestone in Maryland, west of Dan's Mountain.

This group, with its limestone indicated on the map, cropping out on the east flank of Dan's Mountain, dipping under the Potomac and George's Creek coal field. It rises to daylight along the west flank of Savage Mountain and is again found underlying the Meadow Mountain and the Yohiogheny coal fields.

COAL FORMATION NO. 19.

Seral series, Pennsylvania reports.

In the Pennsylvania report this formation is divided into

1. Seral conglomerate, usually known by the name of "millstone grit." In Pennsylvania it exists in the anthracite coal regions of great thickness, made up principally of large pebble of quartz, with a silicious cement. It thins off southwestward, and in Maryland it does not exist; its place being supplied by beds of coarse gray sandstone of moderate thickness.

2. The coal formation proper.

Of this there are four divisions recognized in the Pennsylvania reports, but the necessity of this classification does not seem to me clearly apparent.

The coal formations of this State consist of alternations of

1. Sandstones usually gray and varying from coarse to fine and shaly.

2. Shales usually black and bituminous, or carbonaceous, but also gray, brownish and olive colored.

3. Slate clay, or massive beds of indurated clay, with little or no fissile structure, and known by miners as fire clay.

4. Bituminous coal.

5. Iron ores of the variety called carbonate of iron. This lies within thin strata between the shales and fire clays, or in flattened nodules imbedded therein.

The economic application of the coal and other matters of this formation will be noticed in a subsequent chapter. Its geographic position will be seen by reference to the map which shows that, in this State, it only occurs in Allegany county, and in three separate basins.

FORMATION No. 20.

Mezozoic series of the Pennsylvania reports. New red sandstone of the New York reports and of the British geologists.

We have now enumerated the rocks of our State, beginning with the metamorphic and other of the oldest formations up to the coal. We have found, as a general rule, the older rocks successively covered by those of more recent origin as we proceeded westward. The most marked exception to this is in the case of the new red sandstone of Carroll, Frederick and Montgomery counties, which rest upon the metamorphic and the oldest of the sedimentary rocks.

It consists, at its base, of coarse red and brown sandstones and conglomerates upon which rests finer grained sandstones, followed by red shales of considerable thickness, some of which are calcareous. Upon these shales we find a coarse calcareous conglomerate or breccia of considerable thickness in the southern part of Frederick county.

The older rocks having now been briefly noticed, our attention will be directed to those supposed to have been deposited at much more recent periods, and which are found upon and southeastward of the metamorphic division, (No. 5.)

FORMATION No. 21.

Cretaceous group or chalk period. These do not exist either in Pennsylvania or New York.

It was long supposed by certain geologists that this formation did not exist in this State, or rather that it was covered by the tertiary beds. This view is expressed also in the article of the geology of the United States in that valuable work called "Johnston's Physical Atlas."

Although the present state of our work will not allow me to give its precise limits, yet there is ample testimony to prove that a wide belt of this formation exists in Maryland. It consists of—

1. A thick group of sands and clays of various colors, but principally white, red and blueish gray, with some thin beds of feruginous sandstone resting immediately upon No. 5. In some localities it abounds in lignite derived from coniferous plants. The blueish gray varieties derive their color from

the carbonaceous remains of plants; but we have not yet met with fragments of sufficient size for determination.

Recently, through the patient search of Mr. Uhl, of this city, marine fossils have been found in the white sandy clays near Baltimore, and I have also met with them in a deep cut on the Washington railroad, about twenty-two miles from this city. These fossils are imperfect silicified casts, and have not yet been fully determined but believed to belong to the cretaceous group.

These clays and sands are doubtless the equivalents of those which underlie the green sand of New Jersey.

2. Iron ore clays, (No. 22, in the illustrations). This subdivision consists of a series of beds of fine gray and lead colored clays containing several courses of carbonate of iron in flattened nodules and masses, varying in size from a pound or two to half a ton or more in weight. The color of these clays is due to carbonaceous matter.

The fossils are—

A new genus of a Cycas, of large dimensions, which will be described on another occasion.

Silicified coniferous wood.

Lignites, (coniferous.)

A fragment of a rib of a whale of large size.

A part of the teeth and bones of an herbiferous Saurian, a large extinct reptile allied to the lizard, crocodile, &c.

In the vicinity of Baltimore there are beds of dark grayish colored clays, from which is manufactured the finest brick in the United States. These beds are of moderate thickness and extent. The only fossils are a few fresh water shells, which have not yet been sufficiently investigated to determine the place of this clay in the series. It is more recent than the iron ore clays, (*b*).

The lower green sand, constituting the upper beds of this group, have not yet been sufficiently investigated in this State to permit anything more than a general sketch of them to be presented at this time. The topographical character of the country is such as to present little opportunity for making such geological sections as will give the relative position of the different beds.

So far as can be made out at present, we find them to consist of—

1. Green sand, more or less mixed up with siliceous sand and containing shells.
2. Blueish sandy clays.
3. Black or dark gray micaceous sandy clay, which, in some localities, abounds with fossil sharks teeth.
4. Sandy limestone or indurated marl with numerous shells. This exists in irregular and interrupted beds, from one to four or five feet thick.

5. Beds of reddish and sometimes yellowish sandstones and conglomerates, containing green particles and very few shells. They are from two or three to twenty or thirty feet in thickness.

6. Loose siliceous or common sand, intercalated between some of the beds, but most abundant in the upper portion.

My efforts to collect a complete suit of the fossils of this interesting formation have not been successful, because of its presenting no good natural sections, and there being few or no excavations made therein. Our farmers, except in Cecil county, do not attach the same value to the green sand as is done in New Jersey.

The shells at the outcrops of the beds are usually too soft to bear handling.

Among those already collected are exogyra and belemnites in Cecil county, and cucullea terminalis in Prince George's.

Renewed efforts in this regard will be made hereafter.

This formation will be seen from the map occupies a considerable area within our borders, embracing parts of Cecil, Kent, Harford, Baltimore, Anne Arundel, and Prince George counties.

On the northwest its lowest clays rest upon the metamorphic rocks, (No. 5,) with a very irregular outline, whilst the subjacent rocks may be seen beneath them in the ravines and valleys of the water-courses down to the heads of tide-water, except at the Patuxent.

The southeastern limits where they pass under the tertiary formation are by no means yet determined. It is *provisionally* traced on the map in order to give a general idea of the geographical position of the formation.

The soils of the clays of this formation are variable. Those of the sandy clays with proper treatment are productive, whilst those of the stiff clays are expensive to cultivate.

FORMATION No. 23. TERTIARY.

This series formerly described as one formation has been latterly separated into three, called Eocene, Miocene, and Pliocene. Our work, however, has not progressed so far as to make the separation at the present time. I have, therefore, retained the name tertiary for the three groups.

They consist of nearly horizontal but intercepted strata of sands and sandy clays usually grayish or blueish in color. There are also numerous beds varying in thickness from a few inches to ten or fifteen feet made up in a great measure of marine shells. Corals are also found among them, and in some localities in such abundance as to indicate that they are the remains of elevated coral reefs or islands.

These groups contain some hundreds of species of shells,

which with the corals must have existed when the climate of our latitudes was much warmer than at the present time.

Besides these shells and corals, it also contains remains of whales, dolphins, sharks, and other fish, to be hereafter noticed.

The tertiary groups occupy all of the western shore south of the cretaceous, including St. Mary's, Charles, and Calvert, and portions of Prince George's and Anne Arundel counties. On the Eastern Shore they are believed to reach from a line a little northwest of Chester river southward to a line running from near the head of the Little Choptank, and eastward to the Delaware line.

They embrace a small portion of Kent and Dorchester, and all of Queen Anne's, Talbot, and Caroline.

They possess the highest agricultural interest from the fact that they contain the very extensive deposits of "shell marl," which have so largely contributed to increase the productive value of the land of the middle counties of the Eastern Shore.

These will be fully noticed in a subsequent chapter.

The soils of this region, except the sandy districts, were originally among the most fertile in the country. A long course of improvident agriculture in former days sadly impaired their producing value. Improved systems of farming have, however, restored the fertility of large portions of it. One of the most efficient means has been in the use of shell marl.

FORMATION No. 24. Post Tertiary.

This formation embraces Worcester, Somerset, and the greater portion of Dorchester, county, and consists of beds of loamy clays and sands, which it is believed have not been elevated more than from ten to thirty or forty feet above the tide level. The numerous islands in the Chesapeake bay are also post tertiary.

As it contains very few fossils and perhaps none that will serve certainly to characterize it, the name is applied because of its position being above the tertiary.

The characters of the soil, in connexion with the circumstances under which this region seems to have been formed, indicates that it consists of sediments derived from the water whilst flowing over it from the various formations already noticed. We can detect in the soil and in its subjacent beds matters that *must* have come from points north and west of the North Mountain. Except in some very sandy districts it furnishes a mixed soil whose fertility is readily maintained or improved.

CHAP. III.

CHAPTER IV.

Chemical and Physical Geology, and its relations to Agriculture.

In this chapter we propose to consider the changes which have taken place and are still occurring inr ocks and minerals, and by which soils have been formed.

When we examine a rock jutting out upon a hill-side, especially if the structure be granular, we shall usually find portions of debris (or the results of its decomposition) near its base. If we expose the surface of a rock which has been covered with earth, (not deposited thereon by water,) we shall often find as we dig downwards a portion of mineral matters such as exist in rocks of its kind showing less and less complete disintegration, then a soft crumbling rock, until at length we reach the solid rock itself apparently unchanged. These facts show that in such cases the rock has been slowly altered by atmospheric agency, and that the earthy covering consists of such of its mineral matters as have not been dissolved and carried off by water.

As was stated in Chapter II, the facility with which rocks are disintegrated varies with their structure and chemical contents. For instance, a pure sandstone with a siliceous cement disintegrates very slowly, and the soil produced from it must be shallow and more or less sterile, whilst many varieties containing numerous grains of minerals easily acted upon disintegrate more readily, and produce better soils. Some varieties of granite as well as gneiss abounding in quartz are slowly acted upon, whilst others in which certain kinds of felspar and mica largely predominate, disintegrate more quickly, and give rise to soils of better kinds. And so on with other rocks.

With these examples, by way of illustration, I shall proceed to consider the geological changes which have occurred within the limits of what is now the State of Maryland, in connexion with the physical and chemical forces which have and will continue to act upon the mineral components of the earth, and by which soils are produced.

There are those who might feel an interest in a full investigation of the origin of the geological formations of our State, including their successive formation beginning with the oldest metamorphic and intrusive rocks. This, however, is too wide a field to enter into at this time, and would be scarcely compatible with the object of this report.

The most important agents by which these changes are effected are water, oxygen, carbonic acid, and ammonia, which are among what are termed atmospheric agents. Changes of temperature also produce important effects.

Rain water carries down with it the substances above named, and distributes them through the soil, and even into solid rocks. Nearly all rocks contain silica chemically united with alumina, lime, potash, soda, magnesia, or protoxide of iron, forming what are termed silicates, and these are soluble in water containing carbonic acid. Water it is true takes them up in very small proportion, especially of silicate magnesia, but in a long series of years extensive effects are in this way produced.

The oxygen of the air being also carried into the soil and rocks, converts the protoxides of iron into the state of peroxide, which is the cause of the changes of color in many rocks (whose surfaces and sometimes the interior also) assume reddish or yellowish aspects.

The disintegration of rocks lying at or near the surface, and also of soil, is hastened by the expansion and contraction resulting from changes of temperature, especially where cold winters prevail.

The freezing of the absorbed water causes expansions which open seams in the hardest rocks, so as to give more ready access to the percolating water with its carbonic acid and other destroying agents.

By these and other means of minor importance rocks are constantly being acted upon, and parts of their constituents are dissolved and carried into the depths of the earth to reappear in spring water. We often hear of "pure spring water," but none such exists in nature; it all contains alcaline, earthy and metallic salts, though usually in very small proportions.

Rocks which have been deprived of considerable proportions of their constituents, so as to readily crumble, are said to be disintegrated or rotten, and if the rock be free from grains of quartz or other nearly indestructible minerals of sensible size, the operation will be continued until the mass be converted into a stiff clay. This is the case with some amphibolite or hornblende, as noticed in Chapter II. Most of the limestones of Carroll, Frederick, and Washington counties also produce a stiff soil, owing to the absenee of sand or grains of quartz.

The granular limestones of Harford, Baltimore, and Howard vary from almost pure to those containing 10 to 20 per cent. of sand, of quartz with mica, talc, etc., and sometimes even more.

Carbonic acid and water are unceasingly dissolving and removing the lime from the surfaces of limestones, leaving, however the insoluble matters behind to constitute earth or soil. It appears, therefore, that the soil and earth resting above limestone, when not transported thereon by water, consists of the impurities previously existing in the rock. If these were principally grains of quartz or sand, the soil will of course be what we term sandy. If, however, they be of such minerals as have been fully decomposed, the soil will be stiff, as in the case of limestone No. 10, and some other rocks.

Slates and shales when very siliceous, are slowly upon acted by carbonic acid and other atmospheric agents, and produce light soils; but those of a fine texture, as in Middletown valley, produce good stiff soils.

In considering the means by which rocks and mineral masses are decomposed, and their constituents in part transformed into clay, sand, and soil, we have thus far only referred to such as remain where they were produced. Such soils are said to be "in situ," or "in place," by way of distinguishing them from such as have been transported by water to greater or less distances.

We have abundant evidence that in all ages of our world water was evaporated, converted into clouds, and again fell to the earth in the form of rain. We have all noticed that during heavy rains the water flowing from the land into the streams, carries with it earthy matters, or in common language becomes muddy.

In this way the debris of rocks in the form of sand and clay is constantly carried from the higher to lower levels. Portions are often deposited near at hand, whilst the remainder is deposited in our rivers, in the bay, or in the bottom of the ocean.

There was a period when the only portions of what now constitutes the territory of Maryland, which was above the tide level, were the formations in the table numbered from 1 to 7.

These were probably much more elevated, especially along their southeastern border, than at present. Rains fell upon them, and the streams and waters generally flowed westward, (just the reverse of what now takes place.) These carried the debris from this elevated land into the great ocean on the west, upon the bottom of which it was deposited. These deposits, with remains of the animals of the ancient ocean, have been successively elevated by subterranean forces, so that we find as a general rule that the formations as we proceed westward are of generally more recent origin, as is demonstrated by their fossils. The last that was deposited was the coal formation, (No. 19,) with the exception of the new red sandstone No. 20. This last, although newer than the coal, rests upon the oldest of our sedimentary rocks in Carroll and Frederick counties.

It extends from the centre of Virginia, through Maryland, Pennsylvania, and New Jersey, near New Brunswick, and is supposed to occuppy what was once the bed of a large river. In process of time the older formations, from No. 1 to No. 7, sank to a much lower level, whilst the elevation of the newer on the west continued until the mountain ridges have attained the heights of two to three thousand feet, which changed the course of the drainage from west to east.

Since this change took place the debris brought down by rains and streams has been deposited on the southeast of the older rocks, (No. 5.) and successively formed the cretaceous, tertiary, and post tertiary, which have also been elevated above the tide level. Deposits of this kind are still in progress in tide water creeks and rivers, as is well known to many of our people.

In addition to the numerous bars which are being formed in the bay, there are continual additions to the area of lowlands at the water level, especially on the lower Eastern Shore counties, and at the head of tide in almost all our rivers and creeks. It is needless to give instances in proof of facts so obvious to all acquainted with our tide-water counties.

Whether these additions are to be elevated for the use of man can only be determined by the wise Providence who directs them.

Subsequent to the deposit of our newest post tertiary, there appears to have been a period during which excessive rains and floods prevailed, and this was before the land had attained its present elevation. It would seem, in fact, from the existence of water-worn pebbles and small boulders, that the cretaceous and the metamorphic rocks, which underlie it, must have been elevated four to five hundred feet since this drift period.

It is probable that it was during this era that the northern portions of Europe and America were deluged by these floods, supposed to have transported icebergs, bearing the enormous masses of rock boulders, which are strewed over the Northern States. These boulders, gravel and coarse sand cover large portions of the Northern and Eastern States, and are the cause of the original infertility of the regions it covers. This drift deposit reaches as far south as the middle portions of Pennsylvania and Ohio, but with diminished size and number of boulders.

Although the great flow of waters from the North did not reach Maryland, there were the local drifts before adverted to. We know they were local, because we can trace them to their origin in the regions drained by the Susquehanna, Potomac and other rivers. Some of their pebbles were derived from the detritus of rocks southeastward of the South Mountain, whilst others consist of chert and minerals which only exist between that and the Savage or Allegheny Mountains.

This seems to have been the last of the great changes which

gave the existing geological features to this part of the continent, and fitted it for sustaining vegetable and animal life.

Originally all the inorganic matters required for the growth of plants were locked up, so to speak, in solid rock, and it was not until the commencement of the changes that have been described, that the plant began to flourish.

When we take into view the composition and structure of the various geological formations, in connection with the fact that during the upheaval of the rocks the strata are usually upturned so as to present their edges to the surface, we would expect to find many varieties of soil. Not only do those of different districts vary from each other; but in the same farm, even in the same field, we find two or more kinds of soil differing materially from each other. This is strikingly illustrated in those of the metamorphic district, (No. 5.)

The rocks of this region generally dip from 40° to 60°, and consist (as was stated in Chap. III.,) of gneiss, mica, slate and hornblende slate, with the intrusive rocks before described. In some localties there are considerable areas of gneiss, producing a soil of medium quality, when the proportion of quartz is small; but a dry sandy soil where the latter largely predominates. But when, as is the fact in numerous localities, there are intercalations of hornblende slate of greater or less thickness, we have a much better soil. The reason of this is that the gneiss furnishes little else required by the plants besides silica, potash and soda, whilst the hornblende slate adds to these lime and magnesia and even phosphoric acid. It also disintegrates more readily than gneiss.

In the northwestern portions of this range, where the mica slates prevail with intercalations of hornblende, we find differences in the soils from causes nearly similar.

Many illustrations might be given among the different formations in all the upland counties, but it will be better to withhold them until final and minute surveys can be made.

Examples may, however, be referred to within the cretaceous, (No. 21,) as well as in the tertiary and post tertiary.

Some of the lower cretaceous clays are so fine and stiff that water can scarcely pass through them, and they also resist the entrance of air; they become a stiff mud when wet, and crack open when dry. When the bed is thin, with a light sub-soil, they can be drained and made productive, but this is impracticable upon a thick bed of clay except at great expense.

In the same field we find sometimes an outcrop of a bed of this kind adjacent to one of sand, and where the two become somewhat mixed; the soil is more manageable and capable of being made highly productive. Parts of this formation consist of sandy clays with a good clay sub-soil produced by the constant washing down the fine matter through the soil. The disadvantages of these soils are, that they have not beneath them inexhaustible supplies of all the requisites for plants, as in some other formations. They appear to be deficient in phosphates and chlorides as well as in lime, but generally contain alcalies.

The middle cretaceous, or *iron ore clays*, (No. 22,) are usually too stiff for profitable cultivation, but when mixed with the sandy strata are productive, owing to the presence of alcalies, sulphurid acid and other matters. They require, however, further examination.

Owing to causes before stated, in Chap. III, the upper beds of the cretaceous, have not yet been much explored in our State. Some of them, although largely composed of siliceous sand, produce fine crops of tobacco, corn and even wheat, owing to the presence of phosphates and other matters, derived from fossils they contain. Others consist of mixtures of siliceous and green sand and marine shells, with intercalated beds of blue and gray sandy clays, some of which contain fossils. These produce some of the most valuable soils in the State, as in the southern part of Cecil and Kent, and in Anne Arundel and Prince George's counties.

Owing to an almost horizontal position of these strata, which have a *very* slight southern dip, the surface boundary between the cretaceous and lower tertiary is very indistinct, and will require a minute survey to determine.

The lower tertiary appears to consist of clays with some fine sands, but the soils they produce are more tenacious than those of the upper cretaceous. In passing from the last, southward in the middle parts of Prince George's and Anne Arundel, and in the northern part of Kent, we first notice these tertiary beds on the highlands with the more sandy beds lying beneath them, the level of which becomes lower and they finally disappear under the lower tertiary. In a single field we find decided differences in the soil, but where they are more or less mixed together by nature, the land has, in some measure, retained its fertility for two hundred years; and this, under a system of cropping, that probably would have completely exhausted any other soil, except in the valley of the Nile, which is annually manured by the overflow of the river.

The beds of the middle and upper tertiary as well as the post tertiary, successibly appear further south and seem to have been made up of the detritus from all the older formations which have been noticed. Whilst these were being deposited beneath the salt ocean, marine animals flourished in great abundance, whose remains of shells, corals, etc., constitute the valuable shell marls of several of our tidewater counties.

The soils of these counties also sustained an unexampled system of hard cropping, which, in years gone by, seriously lessened their productiveness.

By the adoption, however, of improved systems, including the use of the shell marl in some of the counties and other manures, their original fertility has been nearly restored to most of them.

The post tertiary which occupies most of the three lower counties and all the islands of the Chesapeake Bay, was evidently formed from the sediments derived from all the other formations, from the tertiary to the oldest.

As the currents of water by which these earthy matters were transported, lessened in force, they carried little else than the finest materials to these counties. Some of these soils, which, at the first glance resemble clay, are found to consist principally of quartz in the state of an extremely fine sand. But this fine sediment contained the mineral matters required by plants in sufficient abundance to constitute large areas of a fertile soil within these counties.

We have followed the progress of nature through the geological changes by which this part of the continent has been made to assume its present shape.

The "beginning" it is not for the geologist to arrive at by the study of natural causes. We must be content with the best evidence we have in reference to these rocks supposed to be the oldest, and which were formerly termed *primary*. I have indicated a narrow belt of these passing northwestwardly through nearly the centre of the State, and have briefly indicated the mode in which the remaining formations are supposed to have been deposited.

That each of these sedimentary strata was once the floor of an ocean, is proven by the numerous remains of marine shells and crustacea they contain. After being upheaved and they became "dry land," and were acted upon by the physical and chemical agencies before adverted to and converted into sands, clays and soils. All the inorganic or mineral elements required for the growth of plants, were derived from rocks, and must have been supplied to the primeval soils, otherwise plants could not have grown upon them, and our planet would have ever remained a barren waste.

CHAP. IV.

CHAPTER V.

Of the Constituents of Plants.

With the aid of the "illustrations" accompanying this report, I have in the preceding chapters presented such an outline as will, I hope, assist in some degree in advancing our knowledge of the geology of this State.

The principal object in this has been to indicate the means by which soils have been, and continue to be formed, and supplied with the mineral or inorganic elements necessary to the growth of plants.

We have next to investigate the sources and characters of the matters of which plants are composed.

1. The organic elements of plants are the following, viz:

CARBON,	OXYGEN,
NITROGEN,	HYDROGEN.

All of which exist in the atmosphere.

The condition in which they occur is as follows:

1. *Atmospheric air, composed of nitrogen and oxygen.*
2. *Carbonic acid, composed of carbon and oxygen.*
3. *Ammonia, composed of nitrogen and hydrogen.*
4. *Nitric acid, composed of nitrogen and oxygen.*
5. *Sulphuretted hydrogen.*
6. *Carburetted hydrogen.*

The five last named constitute less than 1 per cent. of the air, but yet there is sufficient of carbonic acid, ammonia, and nitric acid, to supply the carbon and nitrogen essential to the vegetable kingdom.

7. *Water*, composed of oxygen and hydrogen, exists in the soil, and as a vapor in the air in variable quantities.

Of the four primary elements above named, with the aid of small proportions of mineral matters, is constituted the whole of what is called the *vegetable kingdom.*

The first plants that grew derived them wholly from the atmosphere, and although it will ever continue to be the principal source, yet a large proportion is furnished by the remains of vegetable and animal matters in the soil.

It is believed that the carbon of plants is wholly obtained from carbonic acid absorbed by their leaves and roots, and the nitrogen mainly from ammonia.

The following table gives the average proportion of carbon and nitrogen in one hundred parts of the crops therein named, and also the quantity of ammonia equivalent to the nitrogen they contain. The plants being dried at 212° .

		Carbon.	Nitrogen.	Ammonia equivalnt thereto.
Indian corn, ...	grain,	54.30	1.65	2.00
Do.	fodder,	*	*	*
Wheat, ...	grain,	46.35	2.5	3.03
Do.	straw,	48.48	0.4	0.48
Rye, ...	grain,	46.00	2.0	......
Do	straw,	49.88	0.36	......
Oats, ...	grain,	51.00	1.8	......
Do.	straw,	50.	0.3	......
Tobacco, ...	leaves and stalks,	......	3.36	4.06
Red clover, ...	hay,	47.	1.70	2.06
Potatoes, ...	tubers,	43.50	1.5	1.82
	tops,	45.	.55	.67

* Not determined.

2. *The inorganic or mineral constituents of plants.*

Of the numerous mineral substances in the earth, only the following ten are believed to be essential constituents of plants, viz:

Silica,	Potash,
Phosphoric acid,	Soda,
Sulphuric acid,	Lime,
Carbonic acid,	Magnesia,
Chlorine,	Oxide of iron.

Carbonic acid was enumerated as one of the constituents of the atmosphere, but it also is a constituent of many minerals, such as limestone and other carbonates. It is found in the ashes of plants, but is considered to have been formed therein during the combustion of the vegetable matters.

Alumina and oxide of manganese have been found very rarely in the ashes of plants, and then in such minute proportions that their presence is considered accidental or unimportant.

All of the above have been shown to exist in our minerals, rocks, and soils.

The proportions of them vary not only in each kind, but also in the different parts of the same plant, and during the different periods of its growth. We find in general that potash, lime, silica, phosphoric, and carbonic acid constitute from three-fourths to nine-tenths of plant ash. The remainder usually consists in sulphuric acid, soda, magnesia, and oxide of iron, and chlorine in small proportions.

From the numerous analyses of the ashes of plants recorded by eminent chemists, I select the following, which form the principal crops of Maryland. The proportion of ashes in each plant was determined after drying at the temperature of boiling water (212°) by various chemists.

INDIAN CORN, OR MAIZE.

100 parts of the grain gives 1.2 per cent. ashes.
100 parts of the stalks and leaves gives 3.6 per cent. ashes.
100 parts of the ashes contain:

	Groin.	Stalks and leaves.
Potash	28.37	35.26
Soda	1.74	1.21
Lime	.57	10.53
Magnesia	13.60	5.52
Oxide of iron	.47	2.28
Phosphoric acid	53.70	8.09
Sulphuric acid	......	5.16
Carbonic acid	......	2.87
Chlorine	trace.	1.08
Silica	1.55	28.

WHEAT.

100 parts of the dried grain gives 2 per cent, of ashes.
100 parts of the dried straw gives 4 per cent. of ashes.
100 parts of the ashes contain:

	Grain.	Straw.
Potash	30.02	17.98
Soda	3.82	2.47
Lime	1.15	7.42
Magnesia	13.39	1.94
Oxide of iron	.91	.45
Phosphoric acid	46.79	2.75
Sulphuric acid	......	3.09
Silica	3.89	63.89

RYE.

100 parts of the dried grain gives 2.3 per cent. ashes.
100 parts of the dried straw gives 5.5 per cent. ashes.

100 parts of the ashes contain:

	Grain.	Straw.
Potash	33.83	17.51
Soda	.39	......
Lime	2.61	9.10
Magnesia	12.81	2.40
Oxide of iron	1.04	1.40
Phosphoric acid	39.92	3.80
Sulphuric acid	.17	.80
Silica	9.92	64.50
Chlorine	.14	.29

Oats.

100 parts of the dried grain gives 3.2 per cent. ashes.
100 parts of the dried straw gives 7.24 per cent. ashes.
100 parts of the ashes contain:

	Grain.	Straw.
Potash	21.15	20.88
Soda	......	4.19
Lime	10.28	7.01
Magnesia	7.82	3.79
Oxide of iron	3.85	1.49
Phosphoric acid	50.44	5.07
Sulphuric acid	......	3.35
Chlorine	.51	3.30
Silica	4.40	49.56

Potatoes.

100 parts of dried tubers and roots produce 4.16 per cent. ashes.

100 parts of dried tops produce 15.00 per cent. ashes.

100 pounds of the dried potatoes=300 pounds, as gathered from the ground.

100 pounds of the ashes contain:

	Tubers and roots.	Stems.	Leaves.
Potash	43.18	39.53	19.84
Soda	4.28	15.77	6.02
Lime	1.80	14.85	27.69
Magnesia	3.17	4.10	7.78
Oxide of iron	.44	1.34	4.50
Phosphoric acid	8.61	6.68	13.60
Sulphuric acid	15.24	6.56	6.37
Carbonic acid	18.29	......	......
Silica	1.94	2.56	6.47
Chlorine	3.73	8.61	7.73

Tobacco,

Analyzed by Dr. Jackson. (See Patent Office Report, 1858.)

The samples were from Prince George's county.

The first from rich soil, and the second from much exhausted soil:

	1. From rich soil.		2. From much worn soil.	
	Leaves.	Stalks.	Leaves.	Stalks.
Potash	17.60	40.12	20.32	27.84
Soda	1.40	9.20	4.36	7.28
Lime	22.66	11.48	25.85	23.88
Magnesia	8.	.80	2.00	.40
Oxide of iron and Manganese	2.80	1.40	1.20	1.30
Phosphoric acid	8.50	12.52	7.15	10.28
Sulphuric acid	8.00	2.04	1.52	4.48
Carbonic acid	18.40	16.00	14.80	18.
Chlorine	3.76	2.96	.92	3.12
Silica	8.60	2.40	21 20	3.20
Percentage of ash obtained from each sample, dried .	14.53	9.2	14.76	8.72

Red Clover.

100 parts produce 8.79 of ashes.

100 parts of ashes of the whole plant, except the roots, contain:

Potash	29.96
Soda	3.13
Lime	21.57
Magnesia	8.47
Phosphoric acid	4.09
Sulphuric acid	2.96
Carbonic acid	18.05
Silica	1.95
Chlorine	7.25

TIMOTHY GRASS.

100 parts produce, 5.29 per cent. of ashes.

100 parts of the ashes contain;

Potash,	31.45
Soda,	1.53
Lime,	14.94
Magnesia,	5.30
Phosphoric acid,	11.29
Oxide of iron,	.27
Sulphuric acid,	4.86
Carbonic acid,	4.02
Chlorine,	1.71
Silica,	31.09

Upon knowing the weight of an average crop of each of these plants, we can now ascertain the aggregate quantity of inorganic matters abstracted from the soil by each crop, per annum.

The following gives the weight, per acre, of an assumed crop, with the amount of inorganic matters in each crop of the dried plant.

The first column gives the weight of the crop after being secured, the second after being dried at 212°.

	Weight of crop per acre in pounds.		Ashes per acre in pounds.
	As gathered.	Dried at 212°	
Indian corn, . . . grain,	2250	2000	12
Do. stalks, leaves, &c., or fodder,	9000	8000	288
Wheat, ... grain,	1000	900	18
Do. . . . . straw,	2000	1900	76
Rye, ... grain,	1450	1300	29
Do. . . . . . straw,	4500	4300	236
Oats, ... grain,	2200	2000	64
Do. . . . . . straw,	3500	3200	241
Tobacco, ... leaves,	800	750	108
Do. . . . . . stalks,	600	550	50
Red clover, . . (hay,)	3000	2700	236
Timothy, . . . (hay,)	3000	2700	153
Potatoes, . . (tubers,)	9000	3000	124
Do. . . . . . tops,	3000	1000	150

In the absence of certain information in reference to the yield of parts of these plants, it became necessary to estimate the yield of the straw of the small grain as well as of the corn fodder, potato tops and tobacco stalks, and also the proportion in them when gathered and dried in the air.

It must be borne in mind that the per centage of ashes in each case, was obtained by burning specimens which had been thoroughly dried at the temperature of boiling water, or 212°.

The carbon, nitrogen or ammonia, required for an acre of each crop, may be determined by multiplying the amounts of the dried plants, in the second column of the table, on page 57, by the figures in the last table, cutting off the two right hand figures exclusive of decimals; thus 2000 lbs. of corn per acre, (dried) being multiplied by 54.30 gives 1086 lbs. of carbon in a crop of 2250 lbs., or about forty bushels; in the state it is housed; and 2000 multiplied by 2 gives 40 lbs. for the ammonia required for the same amount of corn.

It will be easy with the aid of the tables to estimate for larger or smaller yields per acre.

Although the facts given in these tables are of the highest importance in their practical application to agriculture, yet it must be admitted that we are still without sufficient information necessary for a full investigation of many important branches of the subject.

The results of analysis, in the tables, were obtained after the maturity of the plant, but we desire to know the composition of the plant in the different stages of its growth.

Experiments have been made by various chemists, proving that the proportion of both, organic and inorganic elements, vary during the growth of the plant. It is necessary, therefore, to determine the exact proportion of each of the constituents of plants at various periods, in order to supply those required for a maximum crop. It was hoped, some years since, that these investigations would be made under the direction of the Agricultural Bureau of the U. S. Patent Office, but so far it appears that little has been done in that quarter.

As the *whole* country is deeply interested in the production of a few plants, which are of more importance than *all* its other sources of wealth, it would seem proper that the work should be done at the expense of the Treasury of the United States.

To perform these analyses in a proper manner will require for each plant the attention of a really competent professional man, during the whole time of its growth. It cannot be expected, therefore, that any one State will bear the expense of the whole work.

CHAP. V.

It is, however, necessary for the proper advancement of agriculture, and if the general government shall finally refuse to move in the matter, it will devolve upon the States.

If an arrangement could be made, by which each State will require that at least one of its most important staples should be thoroughly investigated, I am sure that our own little State will do as much as any of her sisters.

CHAPTER VI.

Of soils and the causes of their exhaustion.

We have to consider soils—

1. In the state that a bountiful Providence furnished them by the means already indicated, and

2. In the state to which they have been reduced by improvident man.

It has been shewn that all the inorganic constituents of plants are contained in our geological formations, and that by natural causes these are decomposed and disintegrated into what we term earth, clays, sand and soil, and that some of these remain where first formed, while others, especially on hillsides, are carried to greater or less distances, by water, and deposited.

The first were termed soils "*in place*" or *in situ*, and the second are named "*transported or drift soils.*"

As I have before stated, the great northern drift did not reach our State, and for that reason the soils on formation No. 5, (gneiss and mica slate, etc.,) and north and west thereof, are classed as soils *in place*, except along streams and in low grounds.

The local drift also formed deposits over large areas in parts of the tide-water counties.

The soils "in place" are as variable in their constituents as the formations they rest on, and from which they were derived. A single farm, nay, a single field, often contains soils of very different constituents. This is more especially the case where the formation consists of thin alternations of several different rocks, as mica slate, gneiss, hornblende slate, etc., inclined at a considerable angle with the horizon.

The strata southeastward of formation No. 5, are nearly level, inclining but slightly to the southward, and as before noticed these beds are made up of materials derived from all the older formations. We should expect, therefore, and actually find, less variety in the soils of these "tide-water regions." They contain, in fact, all the materials removed from the older rocks well mixed together, which have not been dissolved and carried out to the ocean.

The difference in them is in part due to the greater or less force of the currents from which they were deposited. As a general rule, however, they contained every constituent of the older formations. The finely divided state of these materials is such as permit them to be readily prepared for the uses of plants. To this cause we attribute, in part, the original high degree of fertility which our ancestors found in the tide-water counties of Maryland.

The deposits of the local drifts of former days are frequently confounded with soils "*in place*," but the former may usually be distinguished by their containing water-worn pebbles. They are found in small isolated areas in most of the counties bordering on tide-water.

The existence of organic matters in soils has been already referred to, but their importance is such as to require further attention.

When Baron Liebig's writings gave a fresh impetus to agriculture, more than twenty years ago, public attention was strongly directed to the importance of acquiring a full knowledge of the constituents of plants. The Baron's teachings tended to produce the belief that the atmosphere would furnish all the organic matters required for plants, if a full supply of each of the essential mineral constituents exist in or be supplied to the soil.

Being at that time a farmer myself, I was much interested with this view of the subject, and lost no time in putting it to the test of experiments. I prepared a mixture of mineral matters more than equal to that contained in a crop of corn, a crop of wheat and two crops of clover, according to the best analyses we had at that time. They were applied to several varieties of soil, and the same was done by a friend living at some distance, whose soils differed in many respects from mine.

The spaces to which the materials were applied showed little perceptible increase of product.

In the meantime Lawes, Boussingault, and others tried numerous experiments with analogous results. A controversy between Liebig and Lawes sprang up, which continued during some years, and perhaps is not yet concluded. I have not seen the Baron's last work, but according to Dr. Stoeckhardt, Liebig has rather retracted from his original views upon this subject, but without admitting that he has done so.

The numerous practical experiments which have been made to aid in determining correctly a subject of such importance, have largely added to our knowledge, and have materially promoted the advancement of agriculture. If we cannot adopt the "mineral theory" of Liebig in full, we must at least give him full credit for having done more for promoting a knowledge of the true art of culture than any other man.

Before his time agriculture derived little aid from science, and he is entitled to the credit of instituting a new era in its progress.

In reference to the "mineral theory," I may say that in some lands in Europe, cultivated for one or two thousand years, there must have been large amounts of the mineral constituents of plants taken from the soil. Our Maryland soils, however, are not yet in this exhausted state, and are only deficient in one or two constituents. The very fact that lands have been so long cultivated in Europe, without in many cases other manures than those made from them, (with the exception of lime, and within the last one hundred years gypsum,) proves the existence in soils of large supplies of the minerals required.

A few of these, however, are exhausted by crops under constant cultivation more rapidly than they can be prepared by the natural chemical changes in progress.

A careful study of the subjects connected with the "mineral theory" brings us to this conclusion: *that if a soil naturally contains or be supplied with every mineral essential, in such states as to be available to plants, they will flourish to a certain extent.* The atmosphere will supply carbon, hydrogen, oxigen and nitrogen, in sufficient amount *for a normal or natural growth.* But this does not satisfy the wants of man crowded into populous countries, and whose very existence depends upon an *abnormal* or excessive growth of crops.

Experience has, in my opinion, demonstrated that this abnormal growth, or in plain English very *heavy crops cannot be raised during long periods of time without supplying the soil with manures containing at least nitrogen or ammonia.* And we may add, also, that if the soil be constantly cultivated it must be supplied with matters capable of furnishing humus and carbonic acid in order to produce *heavy* crops.

Experiments have been made with artificial soils entirely deprived of animal or vegetable matter, but supplied with all the requisite mineral matters in the proper state. Seed planted therein vegetated, and the plants grew and perfected their seed.

As plants existed before animals, the first plants on the earth must have derived all their organic elements from the air and water. After the death of these plants, and the animals that fed upon them, their remains were returned to the soil to aid in the growth of their successors. In this way soils have been enriched from the commencement of organic life, except where man has occupied and impoverished them by improvident cultivation.

The most exhausting systems of agriculture were practiced in this State, from its earliest settlement, for more than two hundred years. The soils in many of the counties were supposed to be exhausted almost past recovery, and in former years many of our farmers and planters bid adieu to their homesteads, and sought the virgin soils of the South and West.

Improved systems, however, came into use, at first slowly, but soon rapidly prevailed, so that at this time, in some of the counties, the crops are supposed to have doubled in amount with twenty or twenty-five years, and the value of the lands has increased in a still greater ratio.

Before leaving this branch of the subject I shall call attention to the instructive experiments of Prince Salm Horstmar in reference to the inorganic matters of plants.

He planted oats in artificial soils, in each of which one essential inorganic constituent was omitted.

The results were as follows:

Without silica the plant vegetated, but remained small, pale in color, and so weak as to be incapable of supporting itself.

Without lime it produced its second leaf and died.

Without potash and soda it grew to the height of only three inches.

Without magnesia it was also incapable of supporting itself.

Without phosphoric acid it was weak but upright.

Without sulphuric acid it was normal in form, but weak and produced no seed.

In each case the plant doubtless died as soon as it exhausted the mineral matters from the seed, not contained in the artificial soil.

These among other experiments prove that plants will not thrive if any of their essential constituents be absent from the soil.

We are now prepared to ask why it is that lands once highly productive gradually become less so under cultivation, until in many instances the yield of crop did not pay the cost of raising it?

Were they entirely deprived of their elements of fertility, or only in part? If a part only, it is necessary to know which are absent or deficient.

In pointing to the origin and chemical composition of soils in a previous chapter, it was shown that besides alumna (not a constituent of plants) soils contain silica as sand, and chemical combinations of this substance with potash, lime, soda, magnesia and oxide of iron called silicates.

CHAP- VI.

Phosphoric acid occurs mostly as phosphate of lime and phosphate of iron, though generally in small quantity.

The sulphuric acid, which is required in small proportion, is derived principally from sulphuret of iron or iron pyrites. Chlorine combined with sodium, forming common salt, appears to exist in spring water as well as in rocks and soils in extremely small proportion.

It appears, therefore, that next to sand and silicates of alumina, the mineral matters of the soil are mainly made up of the silicates above mentioned, all of whose bases are essential to plants. They are regarded as insoluble in pure water, and but very sparingly soluble in water containing carbonic acid.

There is good reason to believe that these silicates are in sufficient amount in the greater part of the soils of Maryland (except in those few districts which consist almost entirely of sand) to furnish potash, soda, and magnesia for an indefinitely long period with a judicious system of cropping. In addition to silicate of lime, it seems that there is a necessity for carbonate of lime, which being much more soluble than the silicates is more rapidly abstracted from the soil.

By reference to the tables it will be seen that phosphoric acid constitutes about half of the ashes of corn and wheat, the grains we most export; and yet this important element existed in extremely small proportion in most of the original soils of this State. The plants which have in all time flourished upon these soils have withdrawn the phosphates. The decay of the plants, including of course the original forests, left the phosphates and other minerals on and in the soil as well as vast stores of organic matters. In this way nature stored up in the soil the accumulations of thousands of years. We know too well how rapidly our predecessors exhausted them.

In addition to the loss of phosphoric acid and other inorganic matters, the organic matters called vegetable mould or humus, were in a great degree exhausted by almost incessant cultivation in grain and tobacco crops.

This "*vegetable mould* or *humus*" has been investigated by many able chemists, who have given names to the various matters composing it, which need not be stated at this time. It is sufficient to say that by the action of oxygen it furnishes carbonic acid to the roots of plants, and also absorbs and furnishes them with ammonia.

CHAPTER VII.

Improvement of Soils.

In the early days of agricultural chemistry, it was supposed that its most important object was the analysis of soils, and some striking results were occasionally obtained favoring this view of the subject. Occasionally analysis detected the absence of one or more essential constituents and thus indicated the remedy. Our expectations, in this respect, have not been realized; for as knowledge has progressed it has become apparent that, with the advancement of accuracy and minuteness in analysis, the difficulties have rather increased than diminished.

Although, in common with others, I expected much from this branch of research, I have been forced, I may say, reluctantly, to the conclusion that a reliance upon analysis *only* for sure indications of the causes of sterility in soils was delusive and would not hold good in practice. I have repeatedly expressed myself to this effect.

The first professional gentleman with whom I conversed, that fully agreed with me in this, was Prof. J. C. Booth, of Philadelphia. At this time such views prevail generally with chemists and others, who have devoted themselves to investigations connected with this important subject.

Among others, the distinguished Dr. Anderson, Professor of Chemistry in the University of Glasgow, and Chemist to the Agricultural Society of Scotland, has carefully investigated this branch of Agricultural Chemistry. The professor so fully expresses similar conclusions to those I had formed in this regard, that I cannot do better than to give them in his own language. He says:

"It has become more and more obvious that the question of the composition of a soil is one of extreme complexity. We are now convinced that it will be necessary to commence almost *de novo*, and discarding many of the observations hitherto made, endeavor to determine the fundamental principles upon which the fertility of a soil depends.

"It has been found that while in some instances it is possible to predict, with certainty, from analyses, that a particular soil is barren, in numerous others a barren and a fertile soil may approach so closely, in chemical composition, that it is scarcely possible to distinguish them from each other; and so much is this the case that the analyses of a soil must, at the present moment, be considered in many instances as of comparatively little practical value. No doubt practical deductions of importance, may be occasionally drawn from a careful analyses of a soil, but the great majority of those hitherto made, fail to give the desired information. This may, in part, be owing to the imperfect analyses which have too often been made, but it is certainly mainly due to imperfect knowledge of the chemical conditions requisite for fertility; and until these are clearly known, we cannot expect to derive from the analysis of a soil the important conclusions which it ought to, and, at some future time, certainly will yield."

Although I fully agree that in the present state of agricultural chemistry we must not expect too much from analyses, yet I fully agree with him in the belief that there are cases in which "practical deductions of importance may be drawn from the careful analyses of a soil."

I have secured several samples whose analyses, I think, will prove to be of this description, but it was impossible to complete them in time to be noticed in the present report. These analyses will be of the most elaborate character and require the utmost care. The soil from which one of the samples was taken, I have been informed by most reliable testimony, was planted in corn during seventy consecutive years without manure, and since then a rotation of corn, wheat and clover, and sometimes tobacco has been produced without any other manure than gypsum. A sample was also taken from the spot adjacent to the above, which, it is believed, never was in cultivation, but there is no doubt that they were originally alike.

Samples have also been received from other localities which will be analysed with great care, and I hope will assist in enabling us to form correct conclusions upon this interesting subject.

One serious difficulty in the way of analyses of soils is that we cannot, in most cases, determine the exact condition of their constituents. Each of two soils may shew a very similar chemical composition and yet differ widely in their productiveness, because the constituents required for crops in the one are more soluble, or in a state more readily available to the roots of plants, than in the other. Again the physical conditions of soils are known to exercise great influence upon their fertility.

Sometime since, several samples of soil were sent me from a farm in Baltimore county, about which there were some peculiarities that induced me to pay a special visit to the place. I found part of the soil had been produced from a hornblende rock which usually gives a fertile soil, but in this case the rock had a very coarse granular structure, and the soil consisted, for the most part, of these coarse grains not yet sufficiently disintegrated. The only remedy in such a case consists in heavy application of fresh lime, in order to hasten the disintegration of these coarse grains. Now, an analyses of this soil would have shewn a better supply of the requisites for plants than exist in many soils of the vicinity, which are much more productive.

Another and a serious objection arises from the difficulty of procuring a sample whose chemical constitution is precisely similar to any considerable area of soil around it.

In some cases in order to determine, by analyses, the chemical composition of the soils of a farm, we should require almost as many samples as there are acres. There are also special causes which sometimes occasion erroneous results, such as the existence of the remains of animals which have died on and in the soil. Although the decay may have so far progressed as to leave no trace of structure, yet if the sample should chance to be taken where small animals, such as moles, or even a colony of certain insects have died, the result would be a large excess of phosphoric acid united to lime, iron or alumina.

The analyses that agriculture at present requires are those of plants during different stages of their growth, as I have before suggested. We require more light upon this subject.

The more important characteristics of the soil we can acquire by tracing the soil from its origin through all its changes, including those produced by cultivation and removing crops therefrom. It is to this practical view of the subject, I feel it my duty to invite the attention of the farmers and planters of Maryland.

From what has been said upon the exhaustion of soils in the last chapter, it would appear that such as have long been cultivated without manure, must have been, for the most part, deprived of the small proportions of lime and phosphoric acid they originally contained.

Other essentials have been abstracted to some extent by the crops grown upon the land; but a large portion is again restored in the crop which remains, and in the manure of a well regulated farm.

The principal exceptions are with tobacco, of which the leaves are exported, and hay and straw, which, in the vicinity of cities and towns, are often sold. It will be observed, however, that the ashes of tobacco, as well as the hay, contain small proportions of phosphoric acid, although they abound in potash and lime.

The severe cropping of our soils, especially in the older counties, also destroyed the vegetable mould or humus, with its rich stores of ammonia, and which, as long as it lasted, furnished ample supplies or carbonic acid so necessary for a luxuriant growth of plants.

It would appear, then, that in order to restore our fields to such a condition as will give heavy crops we must return to them ample supplies of lime as well as its phosphates. A judicious use of these with such a system of cropping as will largely increase the supply of humus is necessary. If the land remains several years in grass, in each rotation, the amount of humus or organic matters will be largely augmented and give increased supplies, both of carbon and ammonia, provided the grass be consumed at home, and the manure of the farm be carefully saved and distributed.

If, however, farmers and planters will persist in putting their fields in grain or tobacco every other year, or two years out of three, they will find it necessary to supply ammonia and phosphoric acid, or manures which will produce them. But even with these additions *very* heavy crops are not to be raised continuously for a long period from a soil containing an insufficient supply of humus.

The bad effects of hard cropping are so gradually shewn, as often to escape the notice of cultivators, until serious mischief has been done.

The crops begin to fail in amount after sometime, more especially during unfavorable seasons. It is common to charge it all to the season, or to disease, or the ravages of insects; and it is unfortunately too true that the farmer's hopes are sometimes blasted by these causes, even upon soils abounding with every requisite. It is equally true, however, almost always, that in the most exhausted soils the yield of the crop is proportionally most impaired by the above named causes.

I have been often asked within the last twenty or thirty years, by agricultural friends, "why it is that in certain districts the cornstalks and wheat straw grow as luxuriantly as as ever, while the yield of grain is much less than in former years." The answer is obvious, that by a long system of hard cropping, a deficiency of phosphoric acid and humus has been produced, and consequently the plant is deprived to a great extent of matters essential to perfecting the grain.

The various substances available as manures, or capable of enriching the soil, will be noticed in subsequent chapters; but it is necessary to refer to the action of some of them in this place.

Lime, in several of its combinations, has been advantageously applied to land from a very remote period. According to Plinius it was in use in Germany at the period of the Roman conquest, in the form of marl.

Its uses are two-fold—

1. It is one of the essential constituents of plants,
2. It acts chemically upon both the organic and the mine-

ral matters in the soils, and liberates such as plants require it in a proper state for their use.

The action of lime upon soils may, in part, be illustrated by referring to a few experiments.

1. If we take, for example, felspar (rich in silicates of potash and soda,) powdered extremely fine, we find that water, even with carbonic acid, will dissolve nothing out of it for a very long period, and that boiling acids dissolve little of these silicates, even though continued for weeks. When the powdered felspar is well mixed with lime and exposed to heat, silicates may be readily dissolved out of it by acids. It the mixture be made as first above, and kept wet for some weeks the same results are obtained though in less amount.

A weighed quantity of soil or clay, say 1000 grs., may be placed upon a filter and distilled water passed through until it has abstracted all the soluble matter. If we mix what remains in the filter, intimately with lime, and let it remain in a moist state during some weeks, and again filter, we obtain an additional proportion of soluble matters, besides some of the lime used

When we dissolve out of vegetable matters, all that water will extract, we can obtain an additional amount by mixing them with lime, which also promotes the elimination of ammonia from organic matters.

Although these effects upon both vegetable and mineral matters are more *rapidly* produced by means of quick lime, yet they are *equally certain* with the carbonate as it exists in marl, chalk, or in lime if longer exposed.

These important facts, simple as they are, sufficiently explain why it is that lime has long been considered necessary to a proper cultivation of the soil.

Its good effects have been abundantly experienced in our State, and in some of its modifications it has been a most important agent in renovating "worn out" lands.

Previous to the introduction of crushed bones and guanos, lime was, in some of the counties, the only material used, (besides gypsum and the manures of the farm). It has been and continues to constitute a valuable resource to the agriculture of Maryland. Like most good things, however, it may be and has been in some cases the indirect cause of ultimate injury to the soil.

If we refer to the mineral constituents of our crops we find that lime, in a state of purity, furnishes but one of these constituents, and yet by its action on the components of the soil it is incessantly eliminating others equally necessary.

It will therefore happen in the course of time that one or more of these constituents will be exhausted or much diminished in a soil which has been fully supplied with lime and no other manure.

If the grain only be exported and the straw be restored in the manure of the farm, this system may be continued for a very long time, if the soil originally abounded in the mineral constituents of plants; and the same may be said on a tobacco plantation provided the stalks be returned to the fields.

If, however, the whole crop be sold off as hay, straw, etc., the exhaustion of the soil will be much more rapid.

It should be remembered also, that a soil which has been impoverished by these means can only regain its fertility when supplied with the matters so inconsiderately taken from it. It is generally much more expensive to render such soil again productive, than those which have never been limed.

I would, by no means, discourage the use of lime, which ought to be applied if practicable in some form or other, to all soil in cultivation once in about eight years, but it should be judiciously used, and other manures, especially bones and ashes, should be also applied. It was proper to advert to lime upon the chapter upon the improvement of soils, but it will come under our notice again in the chapter upon lime.

CHAPTER VIII.

Lime and Limestone.

In the sketch of the geology of the State given in Chapter III, the more important formations of limestone were pointed out; but it is proposed in this chapter to describe them with reference to their agricultural applications.

In Harford, Baltimore, and Howard counties there are several ranges of the metamorphic limestone, (No. 11.)

These limestones have always a more or less crystalline granular structure. In color they vary from pure white to gray and blueish gray.

The name *alum limestone* is applied to a very pure variety because of its white color and large crystalline grains. The less pure varieties are usually fine grained, and contain small grains of quartz or sand, besides mica, talc, and a few other minerals disseminated therein.

The purest even contain small portions of magnesia; and this earth exists in the different beds in various proportions, and in some of them nearly equals the lime in amount. These last are called dolomites, and are fine grain, and have a more glistering lustre than the purer limestone. They may be distinguished by the fact that dolomite is slowly dissolved, whilst the purer limestone dissolves rapidly in acids with a brisk effervescence.

The second series of limestones are those of the western flank of Parr's ridge in Frederick and Carroll counties, (also No. 11,) as shown on the map. They have usually a fine grain resembting that of Carrara marble, and they vary in color from white to grayish blue. They contain little siliceous matter, and in general but small proportions of magnesia or other impurities. They have sometimes a slaty structure. Near the southern limits of the formation the proportion of magnesia is somewhat larger.

The fourth series comprises the limestone of the Monocacy valley, (No. 10 on map.) It lies in strata dipping at a high angle, and is generally very pure, containing but small proportions of magnesia, quartz, or any other foreign matter. Some of its beds, however, which have a slaty structure, contain more or less silica, alumina, and oxide of iron.

On portions of the western side of the same valley I have already noticed the calcareous conglomerate, which constitutes the upper member of the new red sandstone formation, (No. 20.) This is also ranked among the limestones, but is little used for lime, because of its proximity to the better material last referred to.

Middletown valley appears to be destitute of limestone except to a small extent near its southern limits, but whether it exists there in available quantity I have not yet been able to ascertain. A careful survey of that region will be required to determine the fact.

The great limestone deposit of the Hagerstown valley also belongs to No. 10.

The strata there have been much disturbed by forces acting from beneath, sometimes highly inclined in one direction, and then the dip reversed. Many of its beds have a wavelike stratification, and others are much contorted.

Its colors are blueish gray, and various shades of blue; some of which are almost black. Along its eastern border, near the foot of the South Mountain, it is more or less slaty, and mixed with silica, alumina, and oxide of iron; but with the exception of this narrow belt, the stone of this formation contains but little foreign matter, except magnesia, in proportions varying from one or two to twenty, and even thirty per cent. in some of the strata,

The ranges of these limestones are indicated on the map and sections.

Progressing westward we next meet with the limestone No. 15*a*, near the western base of the North Mountain. Its geological position and the localities within which it is available, were shown in Chapter III, and it constitutes the main reliance for lime between the North Mountain and Savage Mountain.

Some of the strata consist of very pure limestone, and others are highly siliceous, whilst the slaty varieties contain large proportions of silica, alumina, and oxide of iron. It contains less magnesia than the older limestones already noticed.

The limestone bed in Formation No. 18*b* is shown on the map as underlying the three coal fields of the State, and cropping out around them. Some of its strata furnish pure lime, whilst others are very siliceous.

Within the coal regions there are several thin beds of limestone which are co-extensive with the strata adjacent to them. There is but one of these of sufficient thickness to be available for agricultural purposes to any extent. In thickness it varies from six to ten feet, and small portions of it are slaty, yet it produces good agricultural as well as building lime.

Numerous analyses have been made of the several limestones above noticed, many of which have been published in the reports of Professor Ducatel, and those of my predecessor. The results of a number analyzed in my laboratory were communicated to parties interested, and need not be repeated here, as they possess no interest to the public.

The great importance of these formations of limestone to the agricultural interests of our State makes it necessary that they should be investigated in a most minute and detailed manner—a duty which I have not yet an opportunity to execute.

All of them, except the metamorphic, are pretty regularly stratified, and the characters of each stratum are continuous to considerable distances, whilst the strata vary considerably from one another in chemical composition, and in their value to the farmer.

The proper course to pursue is to make a minute survey of them in such manner as will permit the construction of such a number of geological sections as will permit us to place the out-crops upon the large map, and then to determine by analysis the exact chemical constitution and agricultural value of each kind. There is no other means in my opinion by which the advantages of these valuable mineral treasures can be made fully available to the agricultural interests of our State. An analysis of a sample from one stratum will often give a result differing materially from those adjacent. Every farmer should be put into possession of such information as will indicate to him with certainty the exact composition of the limestone, or the lime, he may desire to purchase or use.

The importance of this will be made manifest by a few illustrations.

Pure carbonate of lime or limestone consists of—

Lime - - - - - - -	56 per cent.
Carbonic acid - - - - -	44 do.
	100

1. Although we have no limestones *absolutely* pure, yet I have known some of the alum limestone to contain but one-half per cent. of foreign matters. It may be, however, that a large quantity could not be obtained quite so pure, and it will be safer to assume 2 per cent. of impurities, leaving 98 per cent. of carbonate of lime. In burning into lime 44 per cent. of carbonic acid is driven off, leaving as the product of 100 pounds of limestone—

	54.88	pounds of pure lime,
and of impurities - -	2.00	do.
	56.88	do. of im pure lime.

100 pounds of such lime will, therefore, contain 3.2 or $3\frac{1}{5}$ pounds of impurities.

2. A specimen from Pipe creek, Carroll county, differing much in appearance from the last, was analyzed by Dr. Piggot, my assistant in the laboratory, and contained—

Lime - - - - - - - - -	53.40
Magnesia - - - - - - -	1.13
Carbonic acid - - - - - -	43.17
Oxide of iron and soluble silica - - -	1.90
Insoluble silica - - - - - - -	.40

By deducting the carbonic acid driven off in the kiln, there will remain 53.40 pounds of lime and 3.43 pounds of impurities. 100 pounds of such lime will contain 6.05 pounds of foreign matters.

3. A specimen from Long Green, Baltimore county, analyzed by Professor Mayer for me, contained—

Lime - - - - - - - -	36.73
Magnesia - - - - - - - -	6.25
Iron pyrites and alumina - - - -	3.30
Insoluble residue, quartz, and talc - -	17.75
Carbonic acid - - - - - -	35.97

Deducting the carbonic acid as before, the product of lime will apparently be 64.03 pounds, but the impurities amount to 27.30 pounds, or very nearly 43 per cent!

4. Limestone from Howard county, near the Patuxent river—

Lime - - - - - - - - -	35.27
Magnesia - - - - - - - -	1.76
Insoluble residue, (quartz and mica,) - -	25.17
Oxide and (sulphuret of iron or iron pyrites) alumina and soluble silica - - - - -	8.39
Carbonic acid - - - - - - -	29.41

When the carbonic acid has been expelled in the kiln, there remains of this what is *considered lime*, but in fact a fraction over one-half consists of foreign matters! And nearly the whole of these are such as abound in all soils.

Taking into consideration the large quantity of lime that should be used upon every farm of moderate size, it is obvious that the cost of hauling this great amount of impurities in certain kinds of lime is a serious matter, and should be obviated as far as possible.

In many instances farmers and others possessing wood or other fuel prefer to purchase the stone, (if it does not occur on their own land,) and burn it in their own kilns. To such the summary of the above results in the following table will show the loss of weight in the kiln, the weight of lime produced, and the weight of the foreign matters therein. The quantity assumed being one ton of 2,000 pounds:

	No. 1.	No. 2.	No. 3.	No. 4.
	Pounds.	Pounds.	Ponnds.	Pounds.
Weight of limestone . . .	2,000	2,000	2,000	2,000
Loss of weight in calcining .	863	863	719	588
Gross weight of lime . . .	1,137	1,137	1,281	1,412
Weight of foreign matters .	40	69	546	707
Weight of pure lime in one ton of the stone . . .	1,097	1,068	735	705

The table needs no explanation, and the farmer will plainly see that if this subject be properly investigated, he can frequently be saved a vast amount of the severe expenses of hauling useless sand, etc., as well as be aided in avoiding the use of inferior qualities of lime on his land.

In many districts there are none other than very impure kinds of lime accessible, but so essential is lime that it is certainly much better to use these than none. Many instances have come under my notice in which they have produced beneficial effects of a decided character. In some districts it has been clearly shown that certain kinds of impure lime produced their good effects more quickly than the pure lime, but they are not so lasting, and the liming must be sooner rendered. The causes of this depends of course upon the characters of the foreign matters. When finely divided siliceous matter is disseminated in the stone, silicate of lime (important to plants) is formed by the heat of calcining. Sulphuret of iron exists frequently in limestone, and a portion of the sulphur combines with the lime, and after exposure to the air forms gypsum. Some of the impurities also contain silicate of potash, and probably other useful matters, but further investigation is required relative to them.

There seems to be still some difference of opinion in reference to the effects of lime containing magnesia.

In former days, when large doses (sometimes even 360 bushels of fresh lime) were applied to the acre in England, serious injury resulted in many instances when dolomitic lime freshly burned was used. This was attributed to the fact of this variety containing nearly as much magnesia as lime. I have myself observed a similar result in Baltimore county, in which it took some seven or eight years for a soil to recover, after which it became quite productive.

This has been attributed to the fact that *caustic* magnesia is injurious to plants, and it very slowly regains carbonic acid so as to become mild again.

Another reason is, that a dolomitic lime, when wetted, does not slake, (like the purer kinds, or those containing moderate proportions of magnesia,) but like hydraulic cement, hardens into grains and lumps which remain a very long time in the soil in a caustic state.

When magnesia exists in small proportion it is not injurious, but beneficial to soils in which it is deficient, as in some of the lower counties on the Eastern Shore.

If the dolomitic kind be used, it is better to mix it with earth or muck, so that it may not become a cement as before noticed.

Upon a future occasion, and after certain investigations shall be completed, it will be proper to go fully into the subject of the advantages as well as to point out *all* the special effects of lime in agriculture. The propriety of using lime or marl as a manure, is now almost universally admitted.

It was first applied to the soil in Baltimore county about sixty years ago, and its use slowly extended in that and the adjacent counties. It was not an easy matter twenty-five or thirty years ago to persuade farmers of our western counties, owning what are called limestone lands, to try it, even on a small scale. I often urged parties to it, but they preferred to be governed by theories of their own, to the effect that their "limestone soils could not possibly be deficient in lime." This affords one among many proofs of the danger of theorizing without a sufficient knowledge of the circumstances upon which the theory is based.

A moderate acquaintance with the principles of geology and chemistry would have shown that the *so-called limestone soils* (when *in place* and not covered by transported matter) consist of the *foreign matters* of the limestone, and that in general the *lime has almost entirely been dissolved out of them*, and carried off in the manner indicated in Chapter IV.

During a recent tour in one of these limestone districts I

was pleased to find the use of lime was becoming general, and its value for improving the soil now strongly insisted upon.

In the tide-water counties the use of lime has been extended to an enormous amount, and it has been one of the important means by which the productiveness of those counties has been so greatly increased.

One source from which a large proportion of the lime is obtained for the use of the tide-water counties is at Wrightsville, Pennsylvania, and although out of this State, yet it is of so much importance to our citizens that it is my intention at a very early period to thoroughly investigate the quarries at that place.

One of the principal proprietors of those quarries, Mr. Briscoe, informs me that there are inexhaustible supplies of each of the several varieties, and that they are prepared to furnish our farmers with whichever may be found best suited to our soils. Its general characters are similar to those of the blue limestone No. 10 of Frederick county. It contains moderate proportions of magnesia, silica, and alumina. One advantage before referred to in lime of this kind is in the fact that owing to the fineness of the silica there is more silicate of lime formed in burning than in those kinds whose silica is in coarse sandy grains.

Oyster-shell Lime.

We have two sources of this material in Maryland, one of which is known by the name of *Indian shells*, from the fact of having been left upon the banks of our tide-water rivers and creeks by the Indians, who seem to have had an especial fondness for the oyster, if we are to judge by the quantity of shells remaining.

In some localities where they have lain many hundreds or perhaps thousands of years, exposed to the weather, they are in such a state of disintegration as to be advantageously applied without burning or sifting.

There are besides large masses of shells which have suffered little change except in the loss of nearly all their animal matter, and should be calcined into lime before being applied to the soil. As there is a considerable portion of earthy matters mixed up with them, it is necessary that they should be screened before being calcined. The fine matter which passes through the screen, as it contains phosphates and animal matters, is a valuable manure especially for the garden.

Recent Oyster Shells.

It is estimated that about five millions of oysters are annually received in Baltimore, more than three-fourths of which are opened in that city for home use and for exportation. There are left at least four millions of bushels of shells which are calcined into lime, and used for the soil principally of the tide-water districts.

The recent oyster shell is composed principally of carbonate of lime, with about 1¾ per cent. of phosphate of lime, and a small portion of animal matter. As it looses nearly half its weight in being calcined, the lime itself contains 3¼ to 3½ per cent. of phosphate of lime, and its great value as a manure is in part owing to this fact.

It should be remembered that as oyster-shell lime only weighs about 50 pounds to the bushel, and therefore should be used in nearly double the quantity that would be required of stone-lime, which, when fresh, weighs about 90 pounds per bushel.

CHAPTER IX.

Marl.

The term Marl, in Britain, is applied to beds of clay or other fine earthy matter containing a mixture of carbonate of lime, and which will crumble down or disintegrate, upon being exposed to the weather.

In this country a wider signification has been given to the term, thus the green sand, so much used in New Jersey, is called Jersey marl, blue marl, or green marl, although it is generally nearly free from lime.

The extensive deposits of marine shells, in many of the tide-water counties of Maryland and Virginia, have received the name of Shell marl.

Our attention will be first directed to the green sand.

The lower green sand of our State is one of the upper number of what the geologists term the cretaceous group; implying that it was deposited during the era of the formation of the chalk beds of Europe. Our green sand is precisely similar to that of Europe, but the chalk beds do not exist with us.

The area occupied by the older or cretaceous green sand, in Maryland, has not yet been finally determined, but it exists in considerable amount in the southern part of Cecil and in the northwestern parts of Kent county.

It has been obtained, in many localties, upon the headwaters of the Bohemia and Sassafras rivers, and has been largely used with manifest advantage by farmers in that region. The upper surface of the marl being but a few feet above tide-water, renders the working of the pits frequently troublesome. A careful survey of that district is requisite.

The proportion of green particles is not large, but it abounds it with fossils shells, so that its principal value depends upon the carbonate and phosphate of lime, and a small proportion of silicate of potash. The silicate of iron is not also without its use.

My anxiety to aid in developing the green sand range in Anne Arundel and Prince George's counties, induced me to spend as much time as was consistent with other duties, in exploring that district. I found difficulties in the way, (owing to the distribution of the strata,) which must be overcome by a laborious and careful survey.

I had hoped to have gained some knowledge of the stratification of the region, from the results obtained by boring the artesian well at the Annapolis Naval School, but upon inquiry was surprised to find no record had been kept of the strata.

In the appropriation for improvement of the State House, made during the last session of the Legislature, a provision was made for an artesian well in the State House yard. I looked with great interest upon that branch of the work, because, in addition to the full supply of water, so much needed for the protection of the building from fire, it would have given a geological section of great importance. It was with much regret, therefore, I learned that the appropriation was insufficient, and that the well was not even commenced.

In making a survey, on foot along the line of the Annapolis railroad, I noticed the red sandstones, with the irregular sandy limestone beneath it, near Crownsville, as they appear in New Jersey and in Cecil county, and also beneath these a bed of dark blue sandy clay, at about sixty-five feet above tide-water. As the green sand was to be expected next below this, I visited Round Bay under the hope that I should find it cropping out at some of the bluffs upon its borders. But, to my surprise, it appeared that the strata dipped downward in that direction so that the blue sandy clay, seen on the railroad, presented itself but a few feet above tide-water on Round Bay.

I noticed the same clay at other points southwestward, and on the farm of Richard Hopkins, Esq., near the head of South river, it was seen (with the sandy limestone containing shells) resting upon green sand mixed up with common quartz, sand and shells. The shells were so tender that it was not possible to collect a sufficient number to characterize, with certainty, the age of the formation. Since my visit, the proprietor dug a few feet deeper into the green sand, which, as I had predicted, contained much less common sand than near the upper part of the bed.

It was also noticed, on the Prince George's side of the Patuxent, at several places, at one of which (the farm of Charles C. Hill, Esq.,) a cretaceous fossil, a cucullea, was exposed in the farm ditches.

It is again exposed in a ravine upon the lands of R. W. Brooke, Esq., in the same county, near the southern corner of the District of Columbia, and about three miles east of the Anacostia bridge, over the eastern branch.

Mr. Brooke found a decidedly good effect produced by the application of *only sixty or seventy bushels* to the acre on a portion of his cornfield. The upper four or five feet of the bed was exposed and had doubtless lost much of its lime and potash. Upon being analysed it was found to contain

Carbonate of lime,	26.5	containing	lime	14.9
" magnesia,	2.7	"	magnesia	1.4
Silicates of alumina, iron and potash,	25.4	"	potash	1.3
Sand,	42.4			
Water,	2.			

The proprietor promised to dig into it at a point further down the ravine, so as to reach the lower part of the bed, but upon calling there seven months after my first visit, I regretted to learn that owing to indisposition in his family, no further progress had been made. It is to be hoped that it will be more fully opened, so as to give access to the purer material which may be expected below. Even that now accessible will doubtless prove a valuable manure, especially for the stiff soils of the cretaceous clays which range immediately on the northwest of it.

The result of my preliminary survey of this formation leads to the conclusion—

1st. That the lower green sand rests upon the southern edges of the cretaceous and iron ore clays (21 and 22) and ranges from the last named locality to near the mouth of the Patapsco, on the Western Shore.

2d. That its upper limits rarely rise many feet above the level of tide-water and frequently do not reach that elevation.

3d. It is usually overlaid by the red sandstone and ferugenous and other sands before noticed, so that it is only visible in the banks of ravines and other low situations.

4th. The inequalities of stratification and the thickness of the superincumbent beds are such, that to the present time, the green sand beds, suitable for manure, are only known to crop out at comparatively few localities.

5th. That the lowest portions, which, in New Jersey, are the best, may be expected to be found near their northwestern border resting upon the clays, Nos. 21 and 22.

During the early part of last winter I paid a visit to New Jersey for the purpose of examining the extensive marl pits, near Blackwoodstown, in Camden county. I noticed there a section formed by cutting to a depth of about thirty-six feet, of which the upper sixteen feet, consisting principally of siliceous and ferugenous sand containing a few shells, was thrown aside. That portion used as manure, the remaining twenty feet, consisting principally of pure green sand without shells, and a small proportion of common siliceous sand.

Immense quantities of this have been taken out from this and numerous other pits in this formation, which extends from near Sandy Hook, in a southwest course, to Salem, and through Delaware into Cecil county, Maryland.

It is largely used in New Jersey, and is also exported to Pennsylvania and New York.

During the time of my visit teams were being laden to transport it a distance of twelve miles. The quantity applied was usually twenty tons to the acre, whilst some used as low as ten tons and others thirty tons per acre.

Some parties, I was informed, hauled it eighteen to twenty miles, so well are they satisfied of its value.

The very inclement weather of last winter, during the only time I could spare from other duties to visit New Jersey, prevented me from fully investigating the agricultural value of the green sand. When we find the farmers of that State (after an experience of more than fifty years,) continue its use, and at so large a cost to those residing ten to twenty miles from the marl pits, we may be sure of its advantage to them, but before I can venture to speak fully upon the subject, it is my intention to visit that region again and attempt to determine its value by the experience of those who have extensively used it.

We can well understand why, in the *very* sandy soils of a portion of New Jersey, this material, so rich in silicates of potash and iron, should be useful, but it is by no means, certain, owing to the almost total absence of lime, whether it will be as effective upon most of our soils, although it may be useful to some of them.

While the Jersey green sand usually contains little or no lime, owing to the absence of fossil shells, the Maryland article abounds in shells and will doubtless be found to contain a much larger proportion of phosphoric acid, which, in the Jersey green sand, is rarely equal to one per cent. Unfortunately the lower green sand of Maryland, so far as has come to my knowledge, has not been dug into at any point sufficiently deep to enable us to judge of its characters fully.

I have only been able to inspect the upper portions of the beds in a few places where it has been recently opened, and in these the deeper the bed was penetrated the better was the marl. Besides the larger proportion of common sand near the surface, much of the lime has been dissolved out of the shells, and also the silicate of potash out of the green particles.

Many analyses are on record of the European and Jersey green sands, which need not be stated at the present time. The purer kinds in Jersey, according to Prof. Rogers, who conducted the geological survey of that State, contain about

50 per cent. of silica,
6 to 07 " " " alumina,
21 to 22 " " " oxide of iron,
9 to 12 " " " potash,
7 to 09 " " " water, and

of lime and magnesia seldom more than one per cent of each. In some parts of the formation its value is lessened by more or less common sand being mixed with it.

The lowest of the tertiary formation, called Eocene, is found covering the cretaceous or lowest green sand on the southeast. It contains thick beds of siliceous sand mixed up with green sand and shells, and is largely developed on the banks of Port Tobacco creek, and other Points in Charles and Calvert counties, and appears very little above the tide level on West river and on Herring bay, in Anne Arundel county.

A sample from the farm of Charles J. Pye, Esq., four miles northeastward of Port Tobacco, consists of—

Silica, ...	55.73	
Protoxide of iron, ...	9.45	
Alumina, ...	5.45	
Carbonate of lime, marine shells, .	12.91	lime, 6.67
Potash, ...	2.07	
Water, ...	9.90	
Magnesia, Phosphoric acid, } .	quantity not determined	
Organic matter. . . .	do.	do.

The sample, as in other cases, in our State, was taken from near the surface, and has undoubtedly been deprived by water of large portions of its lime and potash,

Mr. Pye was engaged, at the time of my visit, in taking out this marl for his land, and also promised to sink lower into the bed in search of a purer article.

A sample of the same kind of marl was furnished from the lands of Dr. George, of Prince George's county, consisting of—

Sand, ...	75.72
Organic matter, ...	1.79
Sulphur, ...	trace
Carbonate of lime, ...	5.32
Phosphate of iron, ...	.44
Magnesia, ...	trace
Alumina and oxide of iron, . . .	14.77
Potash, ...	2.53

This specimen contains a large quantity of potash in proportion to its constituents, other than sand. It will doubtless prove useful if applied in large doses to stiff soils.

It is more than probable that a careful survey of that region will bring to light other parts of the same bed containing much less sand.

On the Eastern Shore the Eocene green sand marl occurs along the Chester river, but is in general covered by the shell marl of the middle or Miocene division, of the tertiary formation.

These, with the upper or Pliocene division of the tertiary; embrace what is termed the shell marl of Maryland.

This inestimable gift to the farmer occurs in great quantities in Queen Anne, Talbot and Caroline and in the northern part of Dorchester, on the Eastern Shore. On the Western Shore we find it to underlay St. Mary's, Charles and Calvert, the southern part of Anne Arundel and the southern part of Prince George's county.

I have before me the result of analyses of a large number of samples of these marls from various localities which I made twenty-six years ago. Of these about sixty may be seen in the report of Prof. Ducatel, for 1834. At that period the amount of phosphoric acid was not determined for two reasons. The first was that the absolute necessity for applying phosphates to exhausted soils had not then been fully demonstrated, as it has since been. Secondly, the means afforded by the state of chemical knowledge, twenty-six years ago, gave such unsatisfactory results for determining small proportions of phosphoric acid that I was unwilling to state them in the returns.

It is not necessary to quote those results in full, but I may state that the proportion of carbonate of lime which is made up of shells, and occasionally fossil corals, is very variable, ranging from twenty to over sixty per cent. It would be difficult to assume an average. Of this carbonate of lime, it must be remembered, fifty-six per cent. is lime and forty-four per cent. is carbonic acid. The remaining constituents of the shell marl is principally sand with small proportions of clay, oxide of iron, phosphoric acid and water.

During my recent investigations in counties containing the

marls, I was anxious to secure specimens from as many marl pits as possible, with a view for repeating the analyses for the purpose of determining the proportion of phosphoric acid, but it not being the season for digging I was unable to obtain such as I thought would represent the average of each bed.

For this purpose it is proper to take the samples from the face of a freshly made excavation by cutting out an equal thickness from the top to the bottom of the pit, in several places; and then after mixing the whole thoroughly, take out a portion for the sample to be analyzed.

Several gentlemen promised to procure me samples during the winter, when they intended to get out marl for their lands.

In order that the whole subject of these marls may be fully examined, I would suggest that the work would be materially aided if every one who works a marl pit, during this season, will carefully select a large sample in the manner above described, and also a number of specimens of each kind of shell, tooth, bone or other fossil they may meet with, and after *carefully* packing, send them to me at No. 73 Smith's wharf, or to No. 19 McCulloh street. The locality from whence they came and the name of the sender should always be distinctly stated.

I selected, in person, two samples for analyses from the lands of David Kerr, Esq., situated on both sides of Island creek, an affluent of the Treadhaven creek, in Talbot county. The results are as follows:

	No. 1.	No. 2.
Carbonate of lime	26.52	26.13
Phosphate of lime	2.90	6.67
Iron, magnesia, water, and organic matters .	3.07	11.62
Sand and earth	67.51	55.58

The carbonate of lime in both these marls, especially the last, appears to be derived principally from coral, the remains of which are abundant.

The second is remarkable for the large proportion of phosphoric acid it contains, which, independant of the lime, must constitute it a valuable manure, as will be seen by comparing it with the phosphatic guanoes, whose value in money is known.

The weight of one bushel of marl is usually estimated to be 100 pounds, which in the coral marl No. 2 contains 6⅔ pounds of phosphate of lime. The usual quantity applied to an acre is 300 bushels; which of No. 2 contains 2,000 pounds *of phosphate of lime.*

By reference to the tables of analyses of phosphatic guanoes in a subsequent chapter, it will be seen that the kinds most in esteem may be estimated to contain about 80 per cent. of phosphate of lime. The usual application of these per acre is 300 pounds, containing 240 pounds of phosphate of lime.

Now, 3,600 pounds or 36 bushels of the above No. 2, marl also contains 240 pounds of phosphate of lime, and it would (even if the lime be not taken into the account) to be as effective for its phosphoric acid as 300 pounds of Colombian, Sombrero, or the best of the Mexican and other good phosphatic guanoes. In addition, however, the 36 bushels of marl would supply the soil with 26 per cent., or more than 10 bushels of carbonate of lime to the acre. But we could not expect an equally good effect unless the marl be ground so as to be applied in as fine a powder as the guano.

The grinding, however, would prove too costly, and it will be doubtless better to avoid that expense, especially as we can attain our purpose by using the marl in the larger doses usually applied.

This will contain 2,000 pounds of phosphate of lime, which is the amount in 2,500 pounds of the best phosphatic guanoes, costing from $28 to $30 per ton of 2,240 pounds; and will supply both the phosphate and carbonate of lime required for many years.

Mr. Kerr pointed out a portion of a field upon which this marl had been applied forty years ago, and no manure since; and yet its good effects are still clearly perceptible.

It is believed that no marl with 25 per cent. of shells contains less than 1 per cent. of phosphoric acid, and analysis shows it to range from this proportion to that from Mr. Kerr's farm so rich in phosphate of lime. We find, therefore, with as little as 1 per cent. that 300 bushels contain as much phosphate of lime as 375 pounds of the best phosphatic guano.

The greatly increased productiveness of the soils of Talbot and other Eastern Shore counties is in a great measure owing to the extensive application of marl. Of late years, however, much lime has been used under the supposition that it would answer every purpose as well as the marl; and as some gentlemen told me, "it was less troublesome."

The fashion, however, is destined again to change, and the marl to assume its proper rank in the agricultural system of the tide-water counties.

Some who have neglected it for several years for the use of lime informed me that they intended to reopen their marl pits this winter, and to use this valuable manure liberally.

Among others, Judge W. B. Carmichael, whose father was among the first that extensively used marl in Queen Anne's county, and who has largely used both that and lime on his lands, stated that he had fairly tried them side by side. The results he finds most *decidedly in favor of the marl.*

Whilst investigating this branch of the subject, I inquired of several gentlemen as to the effects of phosphatic guanoes on land to which marl had been applied. The answer in each case was that "no benefit whatever could be observed." Now, this is precisely the result that should have been expected, because an ordinary dressing even of those marls poorest in shells, contains more phosphoric acid than the usual quantity of phosphatic guano applied, and will supply all of this constituent required by crops for many years.

It was represented to me, however, by some parties that in such cases the yield of wheat was much increased by the use of bones and Peruvian or other guanoes containing much ammonia. The cause of this admits of easy explanation, and will be noticed in Chapter X upon guanoes.

An important question to be determined is, "How frequently should marl be applied to land in cultivation?" The testimony upon this subject is so conflicting that I have not yet been able to form a satisfactory conclusion, and must therefore refer to a future report for an investigation into this matter.

It is in my opinion more than probable that of those marls containing 35 to 40 per cent. of shells readily crumbling to powder, a much less quantity than 300 bushels per acre will be more economical, provided due care be taken to spread it uniformly. The richer the marl in shells or carbonate of lime, the less will be required. It has been found, however, that larger doses may be applied to soils abounding in vegeble or organic matters, but if these be very deficient, a heavy application of marl sometimes produces an injurious effect for a year or two.

The credit of being the first in Maryland who used marl as manure belongs to the late John Singleton, Esq., of Talbot county, not the Rev. J. Singleton, as Mr. Ruffin calls him in his Essay on Calcareous Manures. Mr. Singleton noticed marine shells in a bank from which he was digging earth for making a causeway. His first application was made in 1805 with 80 bushels to a small area of land, and the increased production was such as induced him to persevere. Others soon followed his example to such an extent as greatly to increase the products of his own and other counties.

There is no doubt that the importance of these marl deposits will be more fully estimated hereafter, and that Mr. Singleton will be remembered as one of the benefactors of the State.

Shell marl, as before stated, exists in several counties on the Western Shore. It is very abundant in Calvert, St.

Mary's, Charles, and in the southern part of Prince George's county. Its characters are similar to those of the marls which have been extensively used on the Eastern Shore. A large proportion of the most valuable beds in the above named counties are so much elevated above the water level, as to permit them to be worked out very cheaply.

In excavating the marl on the Eastern Shore, owing to the low level of most of the beds, the work is often much hindered by the water collecting in the pits. On the Western Shore this difficulty will not so frequently present itself.

I am at a loss to understand why it is that a manure, which has produced such beneficial results, and is so extensively used on the Eastern shore of Maryland, and in many counties in Virginia, should be wholly neglected in our Western Shore counties.

The soils of those counties (as is almost invariably the case with soils which have been long under cultivation) are certainly deficient in both carbonate and phosphate of lime. And yet I have noticed, in some instances, that parties were buying lime and phosphatic guanos, from abroad, to supply deficiencies which could have been more cheaply obtained in substance at home.

On the Eastern Shore I learned that the digging of marl is done in the winter; and, owing to its low position in most cases, it is thrown from the pits upon land above the water level, and hauled and applied to the soil when convenient. By this means, full employment is given to the hands during the winter, when there is little other active work to be done on the farm. When an additional force is required, the price is one cent per bushel for digging and throwing out of the pit. There, as in the Jersey marl pits, full employment can be had for all the spare hands on the farm; and, in fact, labor is also drawn from elsewhere, because constant employment is given during the winter.

There are situations on both Shores, at which excellent marl can be dug and put on board of vessels, at a cost of from one to one and a quarter cents per bushel. A proprietor could, in fact, with proper conveniences, do a good business at some of these places, by delivering it on board of vessels for two cents per bushel, including a fair profit; and if the freight be two cents, the marl could be delivered to points not exceeding—say fifty or sixty miles, for four cents, and probably to any part of our bay or navigable rivers for an additional cent.

A few of these localities on each shore may be named. The first is in Talbot county, on the west bank of the Choptank river, thirty miles above its mouth, and five miles from Easton. The second is in St. Mary's county, on the farm of

Joseph Forest, Esq., situated on the west bank of the Patuxent river, about fifteen miles above its mouth. In both of these localities there are unlimited supplies of marl, one-half or more of which is probably carbonate of lime, and it could be delivered on board of vessels (if proper arrangements be made) at very small cost.

If some of those, whose enterprise and energy have so much contributed to furnish the immense quantities of lime used in our tide-water districts, would turn their attention to getting out and selling these marls, their exertions could not fail to be very soon properly appreciated and rewarded. Our shell marls are, in my opinion, *worth more* than the Jersey green sand, which is purchased by farmers residing at points distant from the pits, at a cost (including transportation) of 10 to 12 cents per bushel. It has, in fact, a commercial value, and is exported by railroads as well as by water, to the adjacent parts of Pennsylvania and New York.

If a large trade could be sustained in our marls, it could be taken out at a very low cost by means of the steam excavators, one of which proved so efficient in opening and grading the Northern Avenue, in Baltimore, a few years since.

There is some difficulty in making a comparative estimate of the value of marl, and the lime now supplied to our tide-water districts.

The different varieties of lime we get, when *not slaked*, contains from 35 to 75 per cent. of pure lime, the balance being made up of sand, magnesia, oxide of iron and alumina. If the lime be slaked or long exposed, the proportion of pure lime will be much less, because of the water and carbonic acid absorbed. It will probably be fair to assume (if we regard only the lime) that one bushel of lime will contain as much as 2½ bushels of good marl, and to the latter we should add the value of the phosphoric acid, which should be applied to all soils long in cultivation.

This subject is of paramount importance to our agricultural community, and consequently to the State at large, and I regret that in arranging my programme of duties, more time could not be allotted to it. Whilst I have been able to present some of its interesting features, there are abundant incentives to pursue the study in a thorough manner. In addition to the lime in our marls, the phosphoric acid assumes much importance. As it exists in every shell, and is subject to chemical solutions and depositions, it is, under known natural laws, carried away from certain points and deposited in others.

It is also liable to occur in certain localities from other causes, of which I am reminded by the occurrence of the coral marls in Talbot county. An instance may be cited in the immense deposit of phosphate of lime in the small Island of Sombrero, which lies in the Atlantic Ocean, eastward of Porto Rico. The material is brought here and sold under the name of Sombrero guano. The term guano was originally applied in Peru to the excrements of birds deposited upon the islands along the coast of that country. As rain seldom or never falls upon them, the ammonia formed by the putrefaction of these excrements is, for the most part, retained in combination with acids and mixed with the phosphates and other matters. In other regions the rains wash out the ammonia and other easily soluble substances, leaving behind a mass in which phosphates predominate, and which are therefore called "phosphatic guanoes." The remains of birds in them indicate clearly their origin.

No such remains, however, have been seen in the material brought from Sombrero, as I am informed. On the contrary, all the organic remains which hitherto observed therein are those of *marine* animals. The only way of accounting for the existence of this vast mass of material, containing nearly 80 per cent. of phosphate of lime, is by supposing that it is made up of the remains of the bones of fish and other aquatic animals which, died at the spot or were drifted thereon.

It is well known that coral reefs and islands, especially the latter, teem with animal life. In the case of Sombrero, the island, like others in the vicinity, was raised above the level of the ocean, and in process of time, nearly all the remains of animal life were carried off by rains or passed into the air, except the phosphate of lime, which remained and hardened into a solid rock.

The characters of the many varieties of shells in our cretaceous and tertiary strata are such as to prove they could only have lived in the salt waters of the ocean. The coral indicates not only this conclusion, but also that a climate prevailed during the deposition of the tertiary formations very similar to that of the present tropical regions, which abound with marine animals.

The phosphatic coral marl, on the estate of Mr. D. Kerr, certainly contains more phosphate of lime than belonged to the coral alone, and the surplus must have been derived from the remains of the animals that existed in that region before it emerged above the ocean's waters. The sum of our knowledge upon this highly interesting subject gives strong ground for the hope that we may possess deposits much richer in phosphoric acid. We may find among the results of a minute and systematic survey of our tide water counties, on both shores, deposits, if not as rich as those of Sombrero, yet sufficiently so to render the State independent of the world in this essential element of our agricultural prosperity.

Is it not worth the attempt?

Fresh Water Marl.

This material occurs in numerous places in Washington county, and deserves the attention of farmers in their vicinity.

I had only an opportunity to examine one of those deposits, which is near St. James' College, six miles southward from Hagerstown. It has evidently been the bed of an ancient pond or lake, supplied with water from the limestone spring at the college, which discharges *not less* than 10 or 15 barrels of water per minute, saturated with carbonate of lime.

The barrier at the south end of the pond was gradually worn down, leaving a stream along its length of, perhaps, two miles, which exposes the marl to the depth of from 5 to 7 feet. Its color, when dry, is of a light gray, similar to that of wood ashes; and it abounds with shells, of which I found several species.

Its composition is as follows:

Lime - - - - - - -	52.55
Sand clay and oxide of iron -	6.16
Carbonic acid - - - - -	41.29
Phosphoric acid - - -	Trace.

It appears to be principally composed of carbonate of lime, including a large proportion of fresh water shells, and doubtless contains a proportion of phosphoric acid that must enhance its value as a manure. Whilst *useless* attempts had been made to raise crops upon it, I regretted that I could hear of no cases in which it had been applied to its proper use as a manure.

This, with other deposits of a similar character, is deserving of the careful investigation which I propose to make of them. In the meantime I hope that some of the good farmers in the vicinity of such deposits will make fair trials of this kind of marl, which can be so cheaply procured, and is likely to answer so good a purpose. It should be applied at the rate of 200 or 300 bushels to the acre, in the mode recommended for shell marl.

CHAPTER X.

On Guano.

Maryland was the pioneer State in the use of guano in this country.

According to my recollection, the first trial of it in the State was by Captain Abel S. Dungan, of a few bags brought by him from Peru, and applied to part of his corn crop. This, I think, was about the year 1832, and soon after the importation of it by the cargo soon commenced.

The good effects of this manure upon exhausted soils, brought it rapidly into use, and the high price it reached, induced enterprizing persons to search for it, especially upon those small coral islands whose only inhabitants are birds. The results of this activity brought into notice the various kinds of guano now sold by the dealers.

All of those that can be had in considerable quantity, are sold in Baltimore, which continues to be the leading guano market in this country.

In estimating the value of any guano to the farmer, we may disregard all its constituents except ammonia and phosphoric acid. The value of some of them depends principally upon their large proportion of ammonia. Others containing no ammonia, are valuable for their phosphoric acids alone. There are some again which contain available proportions of both of these matters so important to the farmer.

We may therefore divide the guanoes now accessible to our agriculturists into three genera or groups.

1.—Ammoniated Guano.

The only species of this kind is Peruvian guano, containing from 7 to 18 per cent. of ammonia.

2.—Phosphatic Guano,

Containing phosphoric acid equal to from 16 to 90 per cent. of phosphate of lime.

Of this the species recognized in the official advertisements of the State Inspector of Guano are—

Mexican,	White Mexican,
Colombian,	Brown Colombian, or
Sombrero,	Nevassa.

3.—Ammonia—Phosphatic Guano.

California,

Containing from 3½ to 10 per cent. ammonia with equally varying proportions of phosphate of lime, or its equivalent in phosphoric acid.

African,

Containing variable quantities of ammonia and phosphate of lime.

Before further considering these guanoes, I beg leave to present a number of results of analysis by different eminent chemists.

The first series are selected from a large number of analysis by my friend, Dr. Piggot, and are contained in the following communication with which he has favored me. There is also one included from my friend Dr. C. Morfit, formerly of Baltimore, but now residing in New York:

Baltimore, Dec. 16th, 1859.

P. T. Tyson, Esq., State Agricultural Chemist:

Dear Sir:—In reply to your inquiries concerning the results of my examination of the different guanoes brought to this market, I submit the following statement:

1.—Brown Mexican Guano.

These guanoes have evidently been formed from the excrements of birds which have lost their ammoniacal salts by evaporation in the hot sun, and by solution in the heavy rains that fall in the region from which they are obtained. The small islands called "Keys," upon which they are found, have often an indistinct basin shape, inclining in every direction towards the centre. In consequence of this form, much of the guano is washed during the rain into hollows where it is gathered.

The color of this guano varies from a pale fawn-yellow to a deep chocolate brown, the difference appearing to be due to the varying quantities of humus present in the different varieties. All the genuine lime guanoes, no matter how dark they may be, leave a pure white ash if thoroughly burned, which sufficiently establishes the organic nature of the coloring matter. I have been told by sea captains who have brought in this dark guano, that it was collected among the thickets of bay cedars which are found upon many of the islands of the Caribbean sea. Most of these brown guanoes also contain notable quantities of fibrous roots, another evidence of the orign of their color. There are, however, dark colored soft guanoes, which owe their tint to the large mixture of iron among them. These contain little or no lime, being made up almost entirely of phosphate of iron. Their contents are very variable, especially in the proportion of phosphoric acid, as will be seen by the analysis which I shall presently quote. The following tables will give an idea of the constitution of this article:

	I.	II.	III.	IV.	V.
Water, ...	32.51	29.28	24.89	13.11	22.98
Organic matter, ...	9.35	12.53	20.93	35.49	11.06
Sand, ...	0.66	0.21	0.12	1.09	trace
Lime, ...	28.56	28.21	26.89	20.86	30.78
Magnesia, ...	....	....	....	4.85	....
Phosphoric acid, ...	16.32	16.68	24.34	16.16	31.22
Chlorine, ...	....	....	....	2.25	....
Carbonic acid and other ingredients not estimated, ...	12.60	13.09	2.84	6.19	3.96
Equivalent of phosphoric acid in bone phosphate of lime, ...	35.36	36.14	52.74	35.01	67.64

Of the above, No. I. represents a cargo imported in August, 1854; No. II. the cargo of the "Susan," which came in the following September; No. III. the cargo of the "Mary," imported in 1855; No. IV. is a sample of a deposit examined to determine its value, and No. V. is the sample of a cargo imported in 1855, and marked by the inspector "Brown Mexican AA," but owing its richness to the presence of lumps of Colombian guano intermixed with it. Of the dark iron guanoes, the following, a sample from a small cluster of islands near the South American coast, called the "Brothers," may be taken as an example.

Water, - - - - - -	16.06
Organic matter, - - -	22.92
Sand, - - - - - -	19.68
Phosphate of lime, - - -	9.01
" " magnesia, -	2.86
" " iron, - -	27.19
Carbonate of lime, - - -	5.95
Not estimated, - - - -	2.33
	100.00

Some of these ferruginous guanoes contain not a trace of lime or magnesia, others are remarkable for the presence of considerable quantities of sulphate of lime or gypsum. An example of this will be found in the following analysis of the sample of a cargo from Patagonia, brought into Norfolk in the sumer of 1858:

Water and organic matter, ...	31.34
Silicates insoluble in hydrochloric acid, . .	34.45
(containing of ammonia—1.31.)	
Lime, ...	5.56
Magnesia, ...	trace
Alumina, ...	13.66
Oxide of iron, ...	3.71
Phosphoric acid, ...	5.51
Sulphuric acid, ...	4.21
Alkaline salts, ...	0.41
	99.85

The combination of the basis and acids is as follows:

Sulphate of lime, ...	7.98
Phosphate of lime, ...	4.85
" " alumina, ...	6.38

Of the insoluble silicates above mentioned, only 10.68 escaped solution in caustic potash, so that a large amount of the silica present is soluble in that reagent.

Another guano somewhat resembling this, but of greater value, was brought here from Soldanha bay, Africa, in 1854. I analyzed a sample of it with the following results:

Water, ...	17.06
Organic matter, ...	7.89
Sand, ...	39.59
Lime, ...	9.56
Phosphoric acid, ...	17.54
Iron, alumina, magnesia, &c., (not estimated,) .	8.46
	100.00

The phosphoric acid is equivalent to bone phosphate of lime, ... 38.10

In another guano of this soft variety, sulphate of lime is present in large quantity, but the phosphate of that earth is more abundant than in the Patagonian above described. Of this variety, the following analysis of a sample from *Portland bay*, Cape Colony, Africa, will furnish a good example:

Water, ...	21.37
Organic matter, ...	10.44
Phosphate of lime, ...	40.24
Sulphate of lime, ...	9.88
Sand, ... , . . .	0.61
Iron, alumina, alkalies, &c., (not estimated,)	17.46

Jarvis island guano is another of these gypseous phosphates as the following analysis of the sample of a cargo imported into New York will show:

Sulphate of lime,	76.72
Phosphate of lime,	16.53
Phosphate of magnesia,	1.65
Sesquioxide of iron,	2.71
Alumina,	0.85
Chloride of potassium,	0.05
Chloride of sodium,	0.67
Sand,	1.22
	99.40

The term "Mexican guano" is not strictly applicable to all the different varieties above described, but I have had no choice in the use of it, all the soft phosphatic guanoes below 65 per cent. of bone phosphate, from whatever source they may have been derived, being known by that name in this market.

White Mexican Guano.

This title was originally given to a very light colored guano, consisting chiefly of porous lumps of low specific gravity, but it has since been applied to all guanoes that exceed 65 per cent. of bone phosphate of lime, which is the highest standard of Brown Mexican. The most characteristic specimens of this variety are those which were formerly brought from Pedro Keys, and the following analysis, made of a lump selected by myself from one of the cargoes, will give a good idea of its composition:

Water,	3.80
Organic matter,	7.10
Lime,	43.91
Magnesia,	trace
Alkaline salts,	0.71
Phosphoric acid,	37.12
Sulphuric acid,	trace
Chlorine,	trace
Sand,	0.11
Iron, alumina, and other substances not estimated,	7.25
	100.00

The calculated amount of bone phosphate of lime 80.53, is so near the sum of the lime and phosphoric acid (81.03) that the guano may be considered a phosphate of lime containing a few impurities.

This analysis, however, must not be supposed to represent the value of actual cargoes of this article. Very few of them reach so high a grade; the average per centage of those I I have analyzed, usually varying between 65 and 75 of bone phosphate. Many of the guanoes recently brought in and sold under this name, are, in reality, common soft guano mixed with fragments of hard guano, which raises their per centage as in the case I have cited in analysis No. 1.

Colombian Guano.

This substance is memorable for the singular errors into which several chemists fell in reference to its constitution. When offered for sale here in 1855, the dealers advertised it as a natural *super-phosphate* of lime, publishing certificates of analysis by several well-known chemists. These, after stating the amount of bone phosphate of lime present, gave estimates of *free phosphoric acid* varying from 5 to 11 per cent. of the whole weight of the guano. The simplest tests were sufficient to show the incorrectness of these statements. For example, the watery solution of the guano was neutral to test paper; is never contained, even after boiling a few minutes, one per cent. of phosphoric acid, and what it did contain was invariably combined with lime. I determined to ascertain, if possible, the cause of this mistake, and in order to this end undertook a minute analysis of the article. My results enabled me to announce its true constitution, and to show that the lime existed as the neutral phosphate, consisting of two atoms of lime, one of water, and one of phosphoric acid, instead of the bone phosphate, which has three atoms of lime and one of phosphoric acid. The error was occasioned by overlooking this. The following table expresses my results:

Moisture	2.34
Organic matter and water combined	8.95
Phosphate of iron	0.35
Phosphate of magnesia	0.61
Lime	38.75
Phosphoric acid	46.22
Sulphuric acid	1.96
Chlorine	free
Sand	0.63
	100.81

Upon calculating the results, some lime will be found over and above the exact amount for the neutral phosphate. This I take to be combined with one of the acids of the humic group, because the substance effervesces with acids after ignition, but not in its original condition. I believe the following table correctly expresses its composition:

Neutral phosphate of lime	87.95
Sulphate of lime	4.21
Lime combined with organic matter	1.47
Organic matter	2.29
Phosphate of iron	0.35
Phosphate of magnesia	0.61
Chlorine	trace
Sand	0.63
Moisture	2.34
	100.85
The phosphoric acid is equivalent to bone phosphate of lime	100.14

After the publication of these results Drs. Higgins and Bickell published a paper on the same subject. They pointed out some interesting facts which had escaped my observation. It will be remembered by every one who has seen this guano in its natural state, that it has a glazed surface of a mammillated form which is a mere shell-covering, of a dark brown compact mass. The latter was the portion I examined. They studied the surface also, and found that in it, the phosphate of lime exists in the form of bone phosphate, while they arrived at the same conclusion with myself in reference to the body of the rock.

The first guano of this kind was found upon the surface of primitive rocks on Monk's island, in the gulf of Maracaibo. Afterwards, it was discovered on El Rogue. A substance closely resembling, but not identical with it, has been recently brought in from El Monita. The cargoes of this guano do not come up to the samples, because they contain a large proportion of the rocks of the islands. Still, they furnish the largest amount of phosphoric acid in a given weight of material, and that, too, in a remarkably soluble form. The only difficulty is, that this variety is scarce.

Sombrero Guano.

This guano is the most abundant and uniform source of phosphoric acid now known to commerce. It is incorrectly called a guano, as it has evidently not originated from the excrements of birds, but is clearly a submarine deposite. I am inclined to think, that it is made up of the bones of marine animals, and the excrements of fish and mollusca that feed among the coral reefs. It is composed of phosphate of lime, mixed with some carbonate of that earth, as well as the phosphates of magnesia, iron, and ammonia. It contains a very small quantity of sand and not much organic matter. It is quarried like a rock, and requires to be ground before it is applied to the soil.

The following analysis will show its constitution and exhibit its remarkable uniformity. They are all cargo samples, Nos. I and II having been imported into this market in the fall 1858, No. III being the sample of a cargo taken by the purchasers as it ran through the mill at Petersburg, Va., in the spring of this year. No. I, I sampled myself, in the storehouse after it had been ground. I have, moreover, another analysis of special samples, which have been either sent on from the island, or selected from the vessels here, but as these are the only *full* analysis of cargoes I have made, they will probably convey a better idea of the average production of the island. A great number of determinations of phosphoric acid and lime in cargo samples give parallel results:

	I.	II.	III.
Water and organic matter, ...	7.07	7.07	12.27
Lime, ...	44.66	41.52	40.27
Magnesia, ...	1.56	0.35	1.07
Alumina, ...	4.97	4.13	4.67 (Alumina and Sesquioxid of iron together)
Sesquioxid of iron, ...	2.03	4.97	
Alkalies, ...	0.81	0.81	
Phosphoric acid, ...	34.65	36.80	35.62
Chlorine, ...	0.35	0.35	
Carbonic acid, ...	2.80	2.92	
Sand, ...	0.69	0.69	0.51
Not estimated, ...			5.59
	99.59	99.59	100.00
Phosphoric acid equivalent to bone phosphate of lime, ...	75.04	79.73	77.75

As additional evidence of the character of this guano, I subjoin a copy of an analysis by Prof. Campbell Morfit, of New York.

Water, - - - - - - - - -		3.52
Sand, - - - - - - - -		.68
Organic matters with lime, - - -		12.33
Chlorid potassium, - - - - - -		.09
Sulphate lime, - - - - - -		.86
Bone ash, { Bone phosphate lime,	67.06	64.67
Bone ash, { " " magnesia,		2.39
Phosphate alumina, - - - - -		3.62
" iron, - - - - - -		1.95
Carbonate lime, - - - - - -		5.34

Silicate potassa and lime, - - - -	.76
Oxide iron, - - - - - -	1.10
Alumina, - - - - - -	3.13
	100.44

The total amount of Phosphoric acid is 34.06
Equivalent to Bone phosphate lime—73.93

Nevassa Guano.

This material is in small red pebbles usually rounded, resembling gravel. I have made but a single full examination of a sample, which gave,

Organic matter and water with some carbonic acid, - - - - - -	12.72
Sand, - - - - - - - -	4.11
Lime, - - - - - - - -	29.66
Alumina and Sesquioxid of iron with a little magnesia, - - - - - -	21.50
Phosphoric acid, - - - - - -	31.66
	99.60.

Phosphoric acid equivalent to bone phosphate of lime, 68.49

The sample from which the above analysis was made, was made perfectly dry.

Iron Guano.

Early in 1857, a large number of rocks, supposed to be identical with the Colombian guano already described, were brought into the United States, mainly through the ports of Baltimore, Philadelphia and New York. I never saw any reason to consider them guanoes, they being chiefly composed of Wavellite, mixed with variable quantities of phosphate and Sesquioxid of iron. The following table expresses their composition. No. 1, is a pale phosphate from Testigos; No. 2, a red rock, from the same island.

	I.	II.
Water,	21.05	16.74
Sesquioxid of iron,	4.85	12.96
Alumina,	22.11	20.91
Soluble matter,	17.55	7.74
Phosphoric acid,	33.65	40.45
Sulphuric "		.02
Chlorine,		.12
Phosphoric acid equivalent to bone phosphate of lime,	72.91	87.64

Both the samples contained fluorine, which was not estimated. These phosphates were deemed here as worthless, because they had not a lime base. I never could see any good reason for this opinion, especially, as I proved by actual experiment, that their phosphoric acid was rapidly dissolved by the alkaline silicates, which are of course present in every soil capable of producing wheat or corn.

El Monita Guano.

Several samples of hard guano have been recently brought in from the island of El Monita. They somewhat resemble genuine Colombian guano in their external surface which is white, smooth and mammillated, but differ from it in their general structure, which is scaly, so that they readily split up in parallel layers. They appear to contain a mixture of the two phosphates of lime in varying proportions. The amount of sulphate of lime differs greatly in different samples. And this difference causes a fluctuation in the percentage of phosphoric acid. The following table expresses the composition of the first sample which I examined:

Water and organic matter, - - -	12.28
Sand, - - - - - - -	0.86
Lime, - - - - - - -	35.06
Magnesia, - - - - - -	2.99
Phosphoric acid, - - - - -	27.78
Sulphuric acid, - - - - -	17.61
Not estimated, - - - - -	3.79
	100.00

I suppose these substances to be combined as follows:

Mixed Phosphates of lime, - - -	46.18
Phosphate of Magnesia, - - -	8.65
Sulphate of lime, - - - - -	35.41
Sand, - - - - - - -	0.86
Organic matter, - - - - -	5.11
Not estimated, - - - - -	3.79
	100.00

Equivalent of phosphoric acid in bone phosphate of lime—60.17.

That these proportions vary materially, will be perceived, when I state, that another sample of a similar rock from the same island, brought in by the same schooner, and examined for phosphoric acid and lime contained of

Lime - - - - - - - -	33 75
Phosphoric acid, - - - - - -	42.83

in the hundred parts.

Equivalent of phosphoric acid in bone phosphate of lime—92.80.

Some stalagmites of a conical form and mammillated surface, were brought in at the same time from the same island. I examined a sample of one of these with the following result:

Water and organic matter, - - -	15.73
Sand, - - - - - - - -	1.34
Lime, - - - - - - - -	36.45
Phosphate of iron, - - - -	2 68
" " magnesia, - - - - -	6.22
Phosphoric acid, combined with lime, -	24.96
Chlorine, - - - - - - -	0.29
Sulphuric acid, - - - - - -	9.15
Not estimated, - - - - - -	3.18
	100.00

Of the ammoniated guanoes I have not made so many analysis as of the phosphatic. I am convinced, however, that the assumed uniformity of Peruvian guano, as imported here, does not rest upon a basis of fact. I have analyzed one sample which contained 18.53 per cent. of ammonia. The last sample I examined gave me only 11.40 per cent. of ammonia, and it certainly presented no appearance of having been damaged on the voyage.

California Guano.

There have been various articles brought in under this name. Usually they are very wet and dark colored, when of course their percentage of ammonia is low. The following table expresses the results of my analysis of 'the first cargo brought to this port, and probably represents the average composition of this material.

Moisture	15.63
Combined water, organic matter, and ammoniacal salts	51.04
(Containing ammonia 10.17.)	
Lime	6.11
Phosphoric acid	11.91
Sand and gravel	3.68
Not estimated	11.63
	100.00

An examination for ammonia of a sample recently sent me gave me only 3.46 per cent. of that alkali. Another sample, which I received about two months since, and of which I made a commercial analysis, gave me the following results:

Water and organic matter	38.45
(Containing of ammonia 6.35.)	
Sand and gravel	16.10
Lime	15.55
Phosphate of magnesia	4.61
Phosphoric acid	16.29
Not estimated	9.00
	100.00

I have examined other ammoniated guanoes from Yucatan, Africa, Patagonia, the Galapagos Islands, and other points, chiefly in the Pacific Ocean, but as they are not articles of commerce here, I shall not trouble you with a description of them.

Very respectfully, yours, etc.,

A. SNOWDEN PIGGOT.

I also avail myself of some of the analysis of Prof. S. W. Johnson, the results of which are just published under the title of "Peat, Muck, and Commercial Manures."

1.—Peruvian Guano.

Analysis of four samples.

	1.	2.	3.	4.
Water . . . . } Organic matter . . . . . }	66.32	65.18		
Water			12.63	12.70
Organic matter			52.27	51.46
*Ammonia potential	5.82	5.95 }	16.03	15.98
" actual	8.93	9.08 }		
Phosphoric acid, soluble in water .	4.69	3.64 }	15.19	14.08
" insoluble " .	10.05	10.50 }		
Sand, etc.	1.69	1.52	2.45	2.66
Phosphate of lime equivalent to the total of phosphoric acid.	(av'ge)	21.28		31.69

[* By the term "potential" ammonia, Prof. Johnson means that which will be produced by further chemical changes, in addition to that already existing.]

California Guano, (Elide Island.)

In Prof. Johnson's book are recorded four analyses. Nos. 1 and 2 by himself, 3 by Dr. David Stewart, of Annapolis, and No. 4 by Dr. Deck, of New York.

	1.	2.	3.	4.
Water	27.34	27.60	18.90	22.64
Organic and volatile matter	39.20	38.75	43.30	43.53
Ammonia	10.00	10.06	9.39	11.46
Phosphoric acid (soluble)	5.07	5.31	11	
" (insoluble)	6.66	6.25		
Sulphuric acid	4,94			
Lime	9.67	9.36		
Potash and a little soda	2.50		9.60	
Sand, etc.	2.50	2.52	4.70	3.24

Prof. Johnson, gives the results of analysis of eight samples of Sombrero guano, which he has recently performed. Four of these were taken from the large pieces as imported, and contained phosphoric acid equal to the following proportions of phosphate of lime, viz., 81.75 pr. ct., 79.88, 76.98, and 75.36 pr ct., an average of 78.5 pr. ct. The remaining had been ground, and were on sale in Hartford and Norwich, Connecticut, and contained respectively, 78.50, 73.21, 68 59, and 68.20 pr. ct. of phosphate of lime, the average being 70.9 pr. ct. The difference, therefore, between the lumps and the ground article, is 8.4 pr. ct., which is attributed principally to the moisture absorbed by the ground article.

Among many chemists in Great Britain, who have paid much attention to guano as well as to its adulterations, I may mention Prof. Nesbit, of London, Doctor Cameron, of Dublin, and Prof. Anderson of Glasgow. It does not appear necessary, however, to quote the results of any of their numerous analysis, as a sufficient number have been stated to show the composition of the unadulterated guanoes accessible to the farmers of Maryland.

It appears that the adulteration of guanoes, especially the Peruvian, is very extensively practiced in Great Britain, and I regret to be obliged to believe that frauds of this kind, are also perpetrated in our own country.

In order to protect our farmers against such impositions, t system of inspection of guano was instituted in our state, and it has doubtless been a means of protection to a considerable extent. But yet, it appears from the testimony of many farmers, that they have had palmed upon them, sometimes, inferior or adulterated guano, with the inspectors mark upon the bags.—Gentlemen have informed me, that boatmen who have brought them Peruvian guano, have offered to furnish them with good new bags, for the guano bags containing the inspector's mark!

Suspecting, however, that they were wanted for dishonest uses, they refused to part with them.

There is a peculiar earth on the southern slope of Hampstead Hill, near the eastern limits of Baltimore, of which I have been informed, large quantities have been, and may still continue to be secretly carted into the city. There being no conceivable honest use for which this material can be brought into the city, and it being very similar in color to Peruvian guano, it was reported to be used to adulterate that article, the mixture being put up and sold in old guano bags containing the inspector's mark! Some months since, the inspector called the attention of the police to the affair, who arrested parties, carting away this earth in guano bags during the night. The arrest was evidently made at an injudicious time, because upon examination the bags were found to contain only the earth. If, however, the parties had been watched until they had taken it to their mixing depot, and completed the crime, they might *possibly* have been properly punished.

During the late season of active field-work, I endeavored to collect for examination samples of guano, ground bones, and artificial fertilizers, which had been purchased and received by my farming friends. Finding but few kinds in their possession, I requested that samples might be forwarded me whenever they shall again purchase.

Among others I got in person a sample of guano from Col. Jno. S. Sellman, of Anne Arundel county, which being sold for Mexican AA, should have contained phosphoric acid equalto 55 per ct., or more of phosphate of lime, and yet the analysis showed but 36 per ct. In this case the Colonel paid for 50 per ct. more phosphate of lime than was implied in the purchase, and if the deficiency had not been discovered, he would have suffered a still greater loss by not applying a proper dose of the phosphate to his soil. How much of this guano was sold and used by farmers, I have no means of knowing.

Samples of other guanoes and fertilizers have recently been received and are under examination.

In using an ammoniated guano we should always mix with it a portion of ground plaster in order to prevent the escape of the ammonia or its carbonate.

I may add also that the experience of those who have several times applied Peruvian guano to the same field, has generally shown that after the second or third application it produces little or no good result unless other manures are also applied. In England the same effects have been observed.

This has been attributed to the large proportion of ready-made ammonia, which tends to promote a vigorous growth of crop and thus rapidly abstracts the essential constituents of the soil, including its phosphoric acid. It is for this reason that a better permanent effect results from mixtures of Peruvian and phosphatic guanos than from the former, when applied alone.

CHAPTER XI.

Bones.

Bones were first used as a manure in Germany, and afterwards, in the year 1771, were introduced into England. Little use, however, was made of them prior to the beginning of the present century, since which period their use has rapidly extended throughout Great Britain.

The high prices of bones in England have drawn, and continue to draw, them from almost every part of the world; even the bones of the soldiers who fell at Waterloo, and at the seige of Sevastopol, have contributed to enrich the soil of Great Britain.

The first bones used for manure in this country, it is believed, were crushed at the establishment of Mr. Wm. Trego, and sold to farmers in Harford and Montgomery counties in the year 1836.

They were sold for some time at 33 to 35 cents per bushel, or about half their present value. The prices in England are about 40 pr. ct. higher than they have yet reached in this country.

When I first applied bones in Harford county, in 1839, the operation was watched with interest by my neighbors, some of whom thought they would prove an extravagant and useless application; and there were those who appeared to have formed theories in reference to manures which ruled bones out of the list, because, as they believed, they were of "too dry a nature."

Their good effect, however, soon became manifest, and the result was to produce heavy crops upon soils which had been long lying idle, after having been rendered sterile by improvident planting and farming of former times.

The use of bones soon extended, and my old neighbors are now perfectly willing to pay double the prices which were then thought extravagant.

Whilst in Harford during May last, I had an opportunity to notice the durable effect of bones which I applied to land from seventeen to twenty years since. All the fields to which they were applied continue to produce heavy crops under the judicious management of the present owner, Mr. Hanway.

There was one field of 10 acres upon which I applied 300 bushels of crushed bones. He enlarged it, and applied 15 bushels to the acre over the whole, but finding the 10 acres which I had manured as above so much more productive than the rest, he applied to the latter (which I had not taken in) 18 or 20 bushels more per acre. He expected, by this means, to equalize the fertility of the whole enlarged field. He informs me, however, that his expectation in this regard was not realized, and he was satisfied would not be until he shall apply another manuring of bones, as he intends to do, to the part upon which I had applied none.

Loudon, Johnston and other writers, inform us that the effects of heavy dressings with bones are clearly shown in England to endure for forty to fifty years.

We shall be prepared to discuss the cause of all this after having described the chemical and physical constitution of bones.

A bone may be described in general terms as a spongy structure, made up in part of a frame-work of phosphate and carbonate of lime, whose interstices are filled with animal matter analogous to gelatine, and a small portion of fat or oil. A piece of bone, long exposed to dilute muriatic acid, will be deprived of its phosphoric acid and other mineral matters, and leave the cartilage or gelatine in nearly original form. If we expose a bone in an open fire until it shall burn white, its form will not be changed, but the animal matter will have been burnt away. If, however, the bone be exposed to heat in a close vessel, all its animal matter, except a portion of the carbon, will be driven off. The remaining carbon, with the earthy matters, constitute what is called animal charcoal, ivory black, or bone black.

We have on record numerous results of analysis of bones of different animals, but the following, which gives the composition of the bones of the ox, will answer our present purpose:

Animal matters analogous to gelatin and albumen, called azotic compounds,	33.30
Phosphate of lime,	55.85
" " magnesia,	2.05
Carbonate of lime,	3.85
Fluate of lime,	2.50
Soda, common salt, &c.,	2.45

The above are the results obtained from a fresh clean piece of bone. Those collected by the bone crushers cannot but have more or less of dirt adhering to them, and, after being crushed, they will absorb a portion of water. This adds to their weight probably about 5 pr. ct., and, of course, lessens the proportion of the other constituents; but it will be safe to assume that 100 lbs. of ground or crushed bones of commerce contain an average amount of gelatine and other azotic compounds, ... 32 lbs.

And phosphate of lime, ...		53 "
Of this last there is phosphoric acid . .	24½	"
And lime, ...	28½	"

The proportion of ammonia produced by the decomposition of the animal matters may be estimated to average about 7 parts of the above 32.

We may, therefore, assume the value of 100 lbs. of crushed bones to consist in:

Ammonia, ...		7 lbs.
Phosphoric acid,	24.5 }	53 "
Lime, . . .	28.5 }	
Carbonate of lime, ...		3 "
Fluate of lime, ...		2¼ "
Phosphate of magnesia, ...		2 "
Soda, muriate of soda, &c., ...		2¼ "

In addition to the above, there are carbonic acid and sulphuretted hydrogen, produced by the decomposition of the animal matters.

It has been stated to me that crushed bones had, in some instances, been adulterated with useless foreign matters, but I have met with no certain evidence of the fact; on the contrary, an examination of a number of samples which farmers had received from several different sources, showed them to be as pure as is practicable with an article of that kind.

There are difficulties in the way of adulterating ground bones, occasioned by the fact that a small addition of foreign matters can be readily detected with a good pocket lens, which every farmer ought to possess.

They are not injured if boiled merely long enough to abstract the grease they contain, but if the boiling be continued until more or less of the gelatine be removed, their value is lessened, because it is from the gelatine that the ammonia is produced. Pure fresh bones should lose from 33 to 37 pr. ct. of their weight, when burned in an open vessel until they become white. But if they have been robbed of part of their gelatine they will lose less weight by burning.

Prof. Johnston, in his Agricultural Chemistry, refers to a

discussion which sprang up some years since, in reference to which of the constituents of bones we are to attribute their value. Sprengel asserted that it was to their phosphates only, and this opinion was favored by Liebig. Others again gave all the credit to the ammonia formed from their animal matter. It would, in my opinion, be a waste of time to give the views of the contestants.

Both sides certainly knew, that all soils which are deficient in phosphoric acid, are rendered more fertile when it is supplied; and it would be certainly difficult to find a field long in cultivation whose productiveness would not be increased by the use of ammonia, provided one or more of the essential elements be not deficient or altogether absent.

It seems strange that such a question could have been raised by distinguished men in the present day, when there is certainly no room to doubt for one moment the efficacy of both phosphoric acid and ammonia as constituents of manure.

Much difference of opinion has prevailed from the first use of bones, as to the best mode of applying them. In Germany it was for a long time the practice to burn them. Whether this was owing to ignorance or the want of bone-crushing mills, we do not know. I believe, however, that this practice has ceased, and that crushed bones are now used in both Germany and in France.

Stoeckhardt, in his Agricultural Chemistry, laments that, owing to the want of appreciation of bones in Germany, they are largely exported to England for manure.

In England they are crushed or ground fine, when they are to be drilled in with turnips seed; but a rather coarser kind is used when sown broadcast.

In this country they are also crushed, but the kind suited for drilling in is not often used, owing to its additional cost.

There are three modes of applying crushed bones to the soil:

1. In the dry state, as purchased.
2. Dissolved in sulphuric acid.
3. Causing an incipient decay, or, more correctly, putrefaction of their animal matter.

If the object is the permanent improvement of the soil, without caring so much about a maximum growth of the first crop, the crushed bones may be applied in the dry state, without any previous preparation. This is the least expensive mode, (1.)

When they are applied for the benefit of only one or two crops, without looking to the permanent improvement of the soil, the phosphate of lime may be made soluble by means of sulphuric acid, or oil of vitriol. (2.)

When the object is to have the bones in such a state as to produce an immediate effect upon the first crop, and which will be continued during many years, it is better to treat them as will be hereafter shown, so as to bring their animal matter into an incipient state of putrefaction, improperly called by some fermentation. (3.)

I have had some experience in the application of dry bones to land, and have also been able to collect the opinions of many who have extensively applied them in this manner. It has the advantage of saving time and labor, but requires a larger dose to produce a given effect upon the *first* crop. Its effects, however, are more lasting, and will continue during a long series of years. This method may answer when the ground is intended to be kept permanently in grass. Gypsum should always be mixed with them in the proportion 1 bushel to 10 of bones.

The system of dissolving in acid, I have been always satisfied, is less advantageous than the putrefactive process, and therefore I have never used the dissolved bones.

In a paper read before the meeting of the British Association, at Dublin, in 1857, Sir J. Murray claims that he was the originator of the practice of using dissoved bones more than forty years ago. Long experience, however, in the use of them has induced him to change his opinion upon that subject, and he now objects to the use of dissolved bones. He states that he finds "the soluble phosphates too soluble; that they melt "too fast, and are carried into the subsoil or pass off into "streams during rains."

He adds that his "present views result from many years "experience," and "that they have been confirmed by a "long series of experiments, carried on for him by the gov-"ernor as well as the gardener of the Richmond (England) "Lunatic Asylum."

The prompt action of dissolved bones upon crops brought them prominently into notice, and induced many farmers to prepare and use them, and, besides, induced a host of parties to prepare them on a large scale to save the farmers the trouble of so disagreeable a process, and not without danger. I am fully convinced that if any one will take the trouble to make proper comparative experiments with dissolved and putrified bones, and notice the results, during five or ten years, they will come to the same conclusion as Sir J. Murray did, who has the candor to acknowledge the errors into which he has led his brother farmers.

The books and periodicals for years past contain numerous directions for dissolving bones, and it is remarkable that they should differ so greatly in the proportions of acid required.

In the Patent Office Report of 1856, Mr. Brown recommended the use of five pounds of sulphuric acid to 100 lbs. of bones, and to compost them with muck.

An article in the Country Gentleman of the 28th October, 1858, by Prof. Gilman, of the Va. Military Institute, refers to an article of Prof. Norton, which recommends 50 or 60 lbs. for whole bones and 25 to 45 lbs. for ground bones, and adds that he (Prof. Gilman) found even 100 lbs. of acid were not sufficient to dissolve 100 lbs. of bones.

The real state of the case is, that if it be desired to dissolve all the phosphates in 100 lbs. of bones, or about two bushels, we must apply 59 lbs. of sulphuric acid, whose specific gravity is 1.85, diluted with three times its weight of water. And to effect a complete solution they must be frequently stirred during three or four weeks. If the bones be whole it will require many months to dissolve all their phosphates.

If it be desired to dissolve a part only, a less proportion of acid may be used. My own opinion is, the less the more economical to the farmer in the long run.

We must not omit to count the cost of applying sulphuric acid to bones, which, of course, will be modified by the proportions used.

Let us first ascertain the cost of effecting a complete solution of the phosphate of lime in bones:

1st. 100 lbs. of ground bones, costing	$1.46
59 " sulphuric acid, (3 cts.)	1.77
We should add for labor and the cost of a vat or tub, which is soon destroyed, freight on acid, &c.	.08
	$3.31

2nd. If we use acid sufficient to dissolve half the bones, the cost will be as follows:

100 lbs. bones, ...	$1.46
30 " sulphuric acid, (3 cts.)	.90
Labor, &c., as before, ...	.08
	$2.44

As a bushel of bones will average in weight 45 lbs., we have to deduct 55 per ct. to get at the cost of one bushel; therefore,

One bushel fully dissolved will cost	$1.49
One bushel half dissolved will cost	1.10

It will be seen, therefore, that by dissolving we much more than double their cost, and if but half dissolved their cost is increased more than two-thirds in amount.

CHAP. XI.

It is true that a smaller quantity will suffice for an immediate effect, which may suit a temporary tenant, but let the land-owner bear in mind that the *whole ultimate benefit* is in proportion to the *weight of bones* applied. It is true the action of the acid upon the carbonate of lime produces a portion of gypsum, but so far as that article is concerned, we can purchase it at less than one-fifth the cost of making it.

When bones or phosphatic guanoes are dissolved in acid it is usual to add absorbend materials, so that it may be made sufficiently dry to admit of being spread. Neither lime nor ashes should be used for this purpose, because it would precipitate the phosphate and neutralize the effect of the sulphuric acid.

Great care should be taken when the acid is poured into the water, which must be done before the bones are added. It must be done very gradually, because it generates heat above the boiling point, and is apt to be thrown in the faces and on the clothes of the workmen.

Sir J. Murray thinks there is much loss by the soluble phosphates being carried off by water; but there is good reason to believe that the cause of their effects being so slight after one or two crops, is more owing to certain known chemical reactions in the soil. Soluble salts of alumina and iron, especially the latter, are never absent from soils, and when a soluble phosphate of lime comes in contact with either of these, the phosphoric acid is precipitated as phosphate of iron or alumina. Both of these, according to Bischoff, are among the most insoluble substances known in water and carbonic acid. But some experiments of Dr. Piggot prove that they are soluble in alcaline silicates.

Whilst it does not seem proper to apply sulphuric acid to bones, yet I think it probable that we may advantageously use either that or muriatic acid in *small* proportion to some of the phosphatic guanos, especially to those containing phosphates of iron and alumina.

It remains now to notice the third mode of preparing bones, which consists in causing putrefaction and decay.

This mode has been evidently coming more into use within a few years past, and we often find directions in the agricultural journals for effecting it, most generally by making them into composts with stable manure or other matter. I have, however, met with nothing in that way that appears likely to answer a better purpose than that practiced by me 19 years ago, after experimenting to some extent. And as inquiries have been made in answer to which I had found it necessary frequently to describe the process, it will now be repeated in full.

Having smoothed over the surface of the ground, (under

a shed, if convenient,) place thereon evenly, a layer of 3 in. of ground bones, and then an even layer of good fine soil or earth, free from stones or sticks. Give a good sprinkling of gypsum over each layer of earth. Another layer of bones is applied upon the layer of earth, and the same alternations are to be repeated with the gypsum until we have four of each bones and earth, and the height of the pile will be 24 inches. As the bones are usually dry, each layer should be well moistened with water or *better with urine*, in order to hasten the process. It is proper to place two or more sticks in the pile reaching to its base, which should be frequently examined by feeling them, in order to judge of the degree of heat produced. If the weather be warm they will begin to heat in a few days, and in a week or two will become hot. When upon taking out the sticks they feel unpleasantly hot, the process should be checked by chopping or spading down the mass from top to bottom, which, if carefully done, mixes the materials well together, and they are ready for spreading.

If the process be commenced during cold weather it may be hastened by placing at the bottom a layer of fresh hors dung about 6 inches thick, and covering the pile with straw or fodder to retain the heat.

There is much testimony in favor of using salt as a manure and it cannot be applied more advantageously than with the bones, because it promotes their solubility. It would be better to place the proper dose of salt with the gypsum upon each layer of the earth.

In reference to the quantity of bones to the acre I may say, that after trying them in quantities from 30 bushels down to 10, I came to the conclusion that 10 bushels to the acre was the most advantageous quantity. I became satisfied also that this quantity, prepared as I have just indicated, and uniformly sown, will be as effective for a year or two as double the quantity applied in the dry state.

Should the soil be dry when wheat ground is dressed with dry bones, and continue so for some time after, but little effect will be produced by them upon the autumn growth.

The effect of the putrified bones will be obvious within a few days after the young wheat appears above the surface. The putrefaction in the first case goes on very slowly; but when the bones have been once heated it will proceed more readily and of course furnish an earlier supply of the much needed ammonia, as well as phosphoric acid.

One great advantage of bones over ammoniated guano arises from the fact that putrefaction and decay have progressed in the latter until nearly all the ammonia which they are capable of yielding has been already formed. And as it is very soluble in water, much of it is rapidly washed off during heavy rains, leaving a portion which is absorbed and retained in the soil. This is going on whenever the ground is wet, so that when the soil is not frozen in winter, the ammonia is passing off and there is no crop growing to appropriate it.

When bones are applied, either dry or in the manner I have suggested, (3,) they are giving out their ammonia as the crops require it, but in cold weather the putrefaction is nearly or quite suspended, according to the temperature, and again resumed in the spring; at first slowly and then rapidly in hot weather, when it is most wanted by the crop.

I have very rarely met with those who have used bones for manure without being satisfied with their effects. Experience has shown, however, that their effects are not so promptly evinced in stiff clay soils as in those of a more porous character. The compactness of very stiff soil prevents sufficient access of air to assist in the decay of the bones. When applied to *very* wet soils the animal matters decompose so slowly as to produce little benefit to crops.

Bone Black or Animal Charcoal.

In former days bullock's blood was largely used in refining sugar, but in the improved modern process very little blood is used. The principal reliance is upon animal charcoal through which the hot syrup is filtered for the purpose of being decolored. It is coarsely crushed or ground and the finer portions and dust sifted out, which would otherwise clog the filtering cloth or pass through with the syrup. After each operation the charcoal is again exposed to heat in closed iron vessels, and the dust, etc., sifted out as before. It is this material that is sold for manure under the name of bone black.

All the animal matter, except a portion of carbon, has been expelled by heat, leaving the carbon with the phosphates and other earthy matters of bones, and is, of coarse, valuable as a manure.

I have been informed that the refineries in Baltimore dispose of their bone black to manufacturers of fertilizers in Philadelphia; the whole amount being annually about half a million of pounds.

A sample which I obtained from Dougherty & Woods, of Baltimore, was analysed by Dr. Piggot, with the following results, viz:

Phosphate of lime, ...	70.10
Phosphate of magnesia, ...	.15
Carbonate of lime, ...	11.85
Charcoal, (animal,)	10.98

Oxide of iron and alumina, ...	3.01
Sand, ...	2.83
Soluble salts, ...	.41
Soluble organic matter, ...	.13

It is to be regretted that this large amount of phosphate of lime should be carried out of our State instead of being used at home. There is no doubt of it being valuable for manure as its constituents clearly indicate, because of the phosphate and carbonate of lime it contains. Its carbon also will prove a source of carbonic acid in the soil.

Cracklins or Greaves.

This material consists of the tissues and other matters remaining after the melting and straining off the fat of animals.

At one establishment in Baltimore (the Butchers' Hide and Tallow Association) there are 100,000 lbs. of this material produced per annum, all of which is sold at one cent per lb. to parties in Philadelphia, to be used in the manufacture of Prussian blue. I have no means of knowing the whole amount produced in Baltimore, but it must be considerable.

Boussingault determined the proportion of nitrogen to be 11.88 per ct., which will produce during the decay of the material more than 14 per ct. of ammonia, or nearly equal to the amount in the best Peruvian guano. It seems, therefore, that it would be worth more than one cent a pound for manure, if it were powdered or otherwise reduced to such a fine state of division as would admit of its being properly mixed with the soil. As it comes from the press, its cakes are about 3 feet square and about 6 inches thick, which are easily transported without being packed. It is in fact almost as solid as wood itself, and will require suitable machinery to bring it into a proper state for manure.

It is but very recently I learned that it was produced in quantities worthy the attention of farmers, but it is my intention to examine further into it as early as practicable.

A mixture of cracklins and the bone black of the sugar refiners would constitute a very valuable manure.

CHAPTER XII.

Marsh Muck and Peat.

This material, so abundant in some of our tide-water districts, has much attracted the attention of farmers in the eastern States, as well as in many parts of Europe.

Its characters vary with the localities where it exists, depending upon its relations with the adjacent dry land and the more or less saltness of the water. If the waters which flow from high lands carry the mud or wash upon the marsh, the muck will abound in earthy matters; but in other situations it is found to be more peaty, and to consist principally of altered remains of the plants from which it was formed. This kind is in fact a sort of spongy peat, and is the best for manure. To this variety I give the name of peaty muck.

Owing to the presence of sulphate of iron or copperas, it has often proven injurious to the soil; but if composted with about one-tenth its bulk of lime, and be permitted to remain one year before being applied to the soil, it has, so far as I have learned, proven useful.

It should be also remembered, that these marshes abound with insects and aquatic animals, and that by the death of these and the decay of their remains, the value of the material is much enhanced.

Where the barn or stables are at no great distance from the material, an excellant plan is to cover the barnyard with it, (say three feet deep,) in the autumn, and to spread the manure from the stable evenly over it. It will also be the better for being trodden down by cattle during the winter, as it will be much enriched by their dung and urine.

It has been stated, when this mode of using the peat or muck is practiced, that the effective value of the mass, (although it be one-half peat,) is equal to the same bulk or weight of manure. When the peaty variety is first dug and thrown out, it is saturated with water, which of course should be allowed to dry off before it is hauled to the barnyard or composted. It then becomes an absorbent of ammonia, and also retains the urine and liquid of the manure, which too many farmers allow to run to waste.

Is has been ascertained that dried fibrous peat possesses disinfecting properties in a high degree, owing to its power of absorbing gasses and vapors. When it is heated in close vessels or in closed clamps, it produces a fine porous charcoal, more effective as a disinfective and decoloring agent than powdered wood charcoal. Some trials go to prove that its decoloring power is even greater than that of powdered bone black.

This matter deserves the attention of sugar refiners and others requiring decoloring materials. The only difficulty in the way seems to be that owing to its fineness—it may too readily pass through the filters.

This condition, however, will be all the better for its use in absorbing ammonia and in deodorizing night soil, which will be noticed on a subsequent page.

Sediment from Fresh Water Ponds.

As information is often desired in reference to deposits of this kind, I present the results of an elaborate analysis performed by Dr. Piggot, of a sample from an old mill-pond on the Hampton estate of Jno. Ridgely, Esq. The pond had existed for about 70 years, and in its lower part the sediment had accumulated to the thickness of 10 feet. The dam having been destroyed by a freshet, its energetic owner determined to restore to his fields this material that had been washed from them.

Its composition is as follows:

Insoluble silica and silicates, . .	71.49
Silica, soluble in potash, . . .	4.34
Oxide of iron, ...	9.26
Alumina, ...	2.32
Carbonate of lime, ...	35
Magnesia, ...	91
Soda and potash, ...	1.95
Phosphoric acid, ...	20
Sulphuric acid, ...	trace
Chlorine, ...	1.22
Humus and humic acid, . . .	3.29
Other organic matters and water, . .	4.15

When I saw the place, more than a year since, the material was being taken out and spread over the adjacent ground about 2 feet thick, and a quantity of lime equal to about one-tenth its bulk was strewed over it. The intention was, after letting lie a year, to apply it to the adjacent fields. Although it contains several important elements of plants in suitable chemical and physical conditions, yet it will be necessary to apply it in very heavy doses to produce decisive effects upon the crops.

As the pond was surrounded with limestone, it is remarkable that so little lime should exist in the sediment.

Another sample, (somewhat analagous to the last,) was sent by Dr. E. H. Pierce, of Queen Anne's county, which contained rather more vegetable matter, but its useful mineral constituents were in still smaller proportion.

Lignite.

This has not been usually ranked with manures, but some facts communicated to the French Academy a year ago have induced me to notice it. If the results of the experiments of M. Millot Brule be confirmed, it will prove valuable for a manure, as well as for protecting plants from insects.

Lignite, although much heavier, resembles charcoal in appearance, and in retaining the form of wood, from which it was derived. Carbon predominates in its composition, but there is a large proportion of soluble matters of vegetable origin, as well as some mineral matters. It contains sulphuret of iron or iron pyrites, which is readily decomposed and oxidated, when exposed to the air, producing copperas and also gypsum, if, as is usually the case, lime be present.

The large proportion of iron pyrites has occasioned the name of sulphur coal to be applied to it in Germany.

In the American Farmer of March last or May, I gave an abstract of the paper of M. Brule, with some remarks of my own in reference to the properties of Lignite. I also indicated its existence in a number of counties, and that it was readily accessible in several deep cuts on the line of the Washington & Baltimore Railroad. It was further stated that if found useful it might be collected, ground and sold at a low price in this State.

Numerous inquiries were made of me, (by those wishing to try it,) as to the means for procuring it, as it had not come into the market.

Mr. William Robinson, lime and guano dealer in Baltimore, having offered to grind up a small portion for gratuitous distribution, I had about a barrel of it sent to him, which has been ground and will be given in quantities of a few pounds to farmers and planters for experimental use. It is unnecessary to repeat the article in the Farmer, but I may state that M. Brule found it most effectual in destroying insects, after trying it in divers ways. In Saxony it is used as a preservative of timber, which is immersed for some time in a bath made up with powdered Lignite stirred up in water.

The chemical composition of Lignite is such that it cannot but prove a manure in the proper dose. What is the proper quantity to apply must be determined by experience. As some of it will produce more sulphate of iron than other samples, and as we should avoid an excess, I would try from five to ten bushels to the acre.

If my agricultural friends will give it a careful trial, both as an insect destroyer and a manure, I will, if it shall prove useful to them, make special examinations of the many localities where it abounds. As it will furnish sulphuric acid, it may answer the purpose of gypsum, besides contributing the humus and soluble salts needed by plants.

It should be tried on tobacco beds as soon as the fly appears, and in fact upon all crops infested by insects. It would be well, also, to try its effects upon the young wheat when threatened by the Hessian fly.

It will be likely to prove useful as a disinfectant of night soil, a manure to be considered in the next chapter.

Night Soil.

When we reflect that the whole of the inorganic matters of animal excrements are derived from the soil, and that a very small proportion of those from man are returned thereto, we have no difficulty in accounting for the necessity that impels us to collect matters which contain them from every quarter.

In China, we are informed that not a particle of either urine or fæces is lost; it is all saved and applied to the soil, which accounts for the fact that the soil of that country continues productive after being cropped for thousands of years. In Belgium and Holland it is saved and applied in a different manner, but they are nearly as successful in avoiding waste.

In France most of this important material is totally neglected, in spite of the warnings of science, and even at the large establishments for making poudrette, in the vicinity of Paris, the material is so unartistically manipulated that the poudrette produced does not contain one-tenth of the most valuable of the contents of the night soil and urine.

In some of the best farms in Great Britian the night soil, with the liquid waste from the kitchen, laundry, &c., are conducted through pipes into tanks and from thence spread upon the land, producing the best results. But in the cities it is nearly all wasted. In London nearly the whole passes into the Thames to pollute its waters.

Whether, as has been suggested, the celebrity of the malt liquors of that metropolis is due to the use of this *anamalized* fluid in its manufacture, it is difficult to determine.

Public attention has of late been much drawn to the subject of properly utilizing night soil and sewerage, both in Europe and in this country, as well for the purpose of promoting public health in cities and towns, as to restore to the soil these valuable matters now wasted.

Some parties in New York recently took out a patent for a combination of certain materials, which being thrown in the proper quantity into a privy sink, will completely destroy the detestible odor without injury to the manure. If this can be done at a sufficiently low cost, the material could be taken out and transported throughout the State without inconvenience to any one, and to the material advantage of agriculture. These parties some time since confided to me their secret, and asked my opinion as to the probability of forming a company to carry on the operation in Baltimore. In reply, I suggested that one of them should come on here and put their plan into practice in one fair case, which I could fully inspect, and that I would be a witness to the exact quantity of each of the ingredients, so as to be sure in regard to the cost. And that if they succeeded in completely destroying the smell, whilst the cost of the poudrette would not exceed its value as a manure, I had not the least doubt that parties here would enter into the business. And I promised also to aid them by calling the attention of enterprizing persons to the subject. I have not since heard from the patentees.

Renewed efforts are also being made by some of the nightmen of this city in this direction, and it is my intention to encourage and to aid them as soon as I can find a little leisure.

I am strongly inclined to the opinion that a mixture of the peaty muck, (either dried or charred,) and gypsum, with the night soil, will answer the purpose; but some experiments are requisite to determine the proportions. The subject is important and every effort should be made to get the material into the condition of a commercial manure.

It does not seem necessary to quote the results of the numerous analysis which have been made of human fæces and urine. It may, however, be interesting to many to know that according to an estimate by Boussingault, the excrements of one human being will produce per annum 14½ bushels of wheat. The common opinion of the Chinese is that it is equivalent to the production of the food required for his support.

Sugar Refiners Scum.

At two establishments in Baltimore, sugar is refined to the amount of forty millions of pounds per annum. In addition to the powdered bone black before noticed, which is sold to manufacturers of fertilizers in Philadelphia, they dispose of another article in still larger amounts to manufacturers in this city. This is the scum, as they term it, which is skimmed off from the boiling syrup and subjected to great pressure to separate it from the syrup.

It is packed in old sugar hogsheads and sold at $1.50 per hhd., containing about 1,200 lbs. each, which is equal to $2.50 per ton of 2,000 lbs. It contains—

Water and organic matter, with very little ammonia, ...	67.45
Sand, ...	10.16
Lime, with a trace of magnesia, . .	8.63
Iron, with some alumina, ...	2.20

Soda 1.56—potash 0.16,	1.72
Sulphuric acid,	7.44
Phosphoric acid,	1.91
Silica soluble,	.34
Chlorine,	.15

According to existing modes of estimating the value of manures, it may be worth five dollars per ton if ground and ready for use.

CHAPTER XIII.

Ashes from Wood and Coal.

The average composition of the ashes from the different kinds of wood used in this State are as follows, according to Prof. Campbell:

Potash	9.3
Soda	2.5
Lime	41.2
Magnesia	6.2
Oxides of iron and manganese	1.6
Sulphuric acid	1.5
Phosphoric acid	4.3
Silica	3.2
Carbonic acid	30.7
Chlorine	0.5

It will be seen that each material of which wood ashes is composed, is also an essential constituent of our ordinary field crops. There is no difficulty therefore in understanding why they are so effective as a manure.

The use of ashes is so well understood by every one, that further remarks upon the subject would be useless.

What are called spent ashes from the soap-makers, are produced by adding 8 or 10 per cent. of lime, and leeching out nearly the whole of the potash and soda. A small portion of these being combined with silica remain in the ashes, which also retains, after being leeched, nearly the whole of the remainder of the constituents of fresh ashes, with an addition to the quantity of lime.

The only difficulty in reference to the use of spent ashes

is that the cost of transportation is increased by the large proportion of water they contain.

Coal Ashes.

Now that coal has become the main reliance for fuel, not only in the cities but also to a considerable extent in the rural districts, there is produced a large quantity of ashes therefrom, amounting to from 5 to 10 per cent. of the coal used.

The composition of the ashes of two samples of coal among others, were determined by the late Prof. W. R. Johnson, to be as follows:

	Anthracite.	Bituminous.
Silica - - - -	50.0	76.0
Alumina - - -	38.9	21.0
Peroxide of iron - -	8.0	2.6
Lime - - - -	2.1	0.0
Magnesia - - -	0.9	0.0

There is some variety in the composition of coal ashes as shown by numerous results of analysis which need not be given. In some cases we find small proportions not exceeding 1 per cent. of potash as an insoluble silicate; in others the proportion of lime is greater which is in part combined with sulphur.

The chemical composition indicates that it is nearly, if not altogether useless as a manure.

The very few cases in which it has been of any service whatever, has been on stiff clays, which it tends to make more porous. I am sure, however, that those wishing to improve the physical structure of a stiff clay, can effect this purpose as well, or better by the use of materials on the farm or its vicinity, at much less cost.

It has been stated to me that sifted coal ashes have been used to adulterate wood ashes, as well as other manures, but so far I have no certain evidences of such fraudulent practices.

If any such are practiced, the detection is very easy with the aid of a common pocket magnifying lens, by which the smallest grains of coke or coal always existing in coal ashes will be exposed.

Gypsum.

There are yet various opinions held in reference to the action of gypsum or plaster of Paris as a manure. It was formerly held by many farmers and others, that its principal office was to absorb moisture from the air and give it out to the plant. Direct experiment, however, proved this opinion to be altogether untenable, because gypsum absorbs much less moisture from the air than the same weight of clay, chalk, and most kinds of soil.

As it had been found to increase the efficacy of composts and putrescent manures, many farmers supposed that it promoted the decay of vegetable matters, but well conducted experiments proved that in the small proportions usually applied, it rather retarded than promoted decay.

Agricultural chemistry at length made such progress as to indicate that certain mineral matters (among which are sulphur or sulphuric lime, the components of gypsum) were essential constituents of plants. It was therefore concluded that the use of gypsum was to furnish these matters.

The fact that ammonia or its carbonate would decompose gypsum and would form sulphate of ammonia, (a non-volatile salt,) suggested to Liebig that the principal effect of gypsum was to absorb ammonia from the atmosphere and from rain water. Boussingault and some others attach little importance to this source of ammonia, because of certain estimates they have made of the quantity of ammonia in the rain-water which falls during the growth of a crop of wheat.

The result of one of the experiments of Boussingault was that about three times more lime was found in a crop of clover to which plaster had been annually applied, than in the part of the field not plastered.

This, with other experiments, go to show that lime is, at least in part, supplied to plants from gypsum. Boussingault's nvestigations seem to prove that lime is more readily obtained by plants from gypsum, owing to its greater solubility, than from carbonate of lime.

We cannot pretend in the present state of agricultural knowledge to state the precise action of plaster in all its phases, and it does not at present seem necessary to record the many facts and experiments made in this connection, or the reasonings founded thereon. It may be useful, however, to state some of the most important purposes served by gypsum which seem to be certainly established.

1. It supplies the crop with both sulphuric acid and lime, whilst this last material, in the form of quicklime and marl, is mainly useful in its action upon the soil and its organic matters as before explained.

2. It absorbs ammonia from the atmosphere, from rain-water, and from the manures and organic matters in the soil, and retains it as sulphate of ammonia for the use of the plant, (unless it be washed off by excessive rains.)

3. Direct experiments have repeatedly proven that the efficacy of stable and other putrescent manures is materially increased by being mixed with gypsum.

I have found that stables and privies may be, in a great measure, deprived of smell by the frequent use of gypsum.

It seems needless to give evidence of this effect of gypsum, which cannot have escaped the notice of every observing farmer. They should remember, however, that it destroys the smell by the production of sulphate of ammonia, which, being retained in this manner, greatly increases the value of the manures.

It is proposed to collect additional facts with reference to this very useful article, and fully discuss its merits upon another occasion.

Soot from Chimneys.

The composition of wood soot, as determined by the late M. Braconnot, of Paris, is as follows:

Ulmic or humic, and analogous to humus	30.0
Azotic matter soluble in water, and containing ammonia	20.0
A bitter organic substance	0.5
Insoluble carbon	3.9
Silica	1.0
Carbonate of lime	14.7
Carbonate of magnesia	trace
Phosphate of lime (containing some iron)	1.5
Sulphate of lime	0.5
Chloride of potassium	0.4
Acetate of potash	4.1
Acetate of lime	5.7
Acetate of magnesia	0.5
Acetate of iron	trace
Acetate of ammonia	0.2
Water	12.5

As these results prove that all the constituents of soot are such as to enter into the composition of the plants, we can readily understand why it is so efficient as a manure. Coal soot differs from the above in being heavier, and in containing more nitrogen and ammonia, and is therefore worth more per bushel.

There is a considerable trade carried on in soot in Europe, where it is carefully saved and applied to the soil.

A very common and beneficial mode of using it, is to apply 20 bushels to the acre upon young wheat and clover. In Flanders it is applied at the rate of 50 to 60 bushels to the acre of colewort, (a non-heading cabbage,) and besides furnishing plant-food, preserves the young plants from insects. In this connection it deserves the attention of our planters and others.

There cannot be less than 50 to 70,000 bushels of soot per annum swept from perhaps 100,000 chimneys in Baltimore alone, little of which is saved for its proper uses.

Alcaline Salts.

Many of these contain the constituents of plants, and are otherwise useful to the soil, but with the exception of common salt, their cost is such as not to permit them to be profitably used as manure in our State. Experiments have been made with many of these salts in Europe, and to some extent in this country, and it has been shown that the nitrate of potash, or saltpetre and nitrate of soda, are highly beneficial to grain crops, but it rarely happens that the excess of crop they produce will bring the cost of the salts. And the same may be said of the sulphates of potash and soda.

Many experiments have also been tried with artificial silicates of lime, potash, and soda, which seem to show that they must be produced at much lower prices than has hitherto been the case, before they can be used by our farmers.

Common Salt or Chloride of Sodium.

It has been clearly proven by carefully conducted experiments, that both chlorine and soda are essential constituents, hence the utility of common salt when applied to soils in which these elements are deficient. It has been found, however, that a small excess is fatal to crops. If the proportion of common salt in a soil exceeds 2 per cent., grasses and grains will cease to grow, and their places will soon be supplied by the salsola and other marine plants.

In the small work of M. J. Pierre, before noticed, is the most complete *resumē* of the use of common salt for manure that I have heretofore met with. He records numerous experiments which have been made with different proportions of salt upon different crops. We gather from these experiments that the most useful dose of salt for an acre of wheat is four to six bushels. The smaller quantity is best adapted to the production of the grain, but when we apply six bushels, there is a considerable increase in the proportion of straw. With more than six bushels, there is a still larger increase of straw, and if the land has been also dressed with stable manure, the stalks of the grain are apt to fall.

English farmers have related to me, that the growth of grass is much promoted by the use of salt, and that the quality of the grass is also impoved. If salt be applied to part of a pasture field, the cattle will browse for years thereon in preference to the part not salted, and they will only resort to the latter when the first has been closely eaten down.

So far as has come to my knowledge, the experience in our upland counties is decidedly in favor of applying salt to the wheat crop, but I am without a sufficient number of facts to indicate those portions of the tide water counties in which it may be usefully applied. In Europe it does not appear to have produced any benefit on lands near the ocean or large area of salt water. We may, therefore, doubt whether it will be useful in several of our southern counties.

If bones be used, an excellent mode of applying salt is to add the salt to the compost of bones recommended in chapter XI.

CHAPTER XIV.

Barn Yard Manure.

The agricultural community are so much accustomed to homilies upon the management of the manures from the animals of their farms, that I really doubted the propriety of touching upon the subject. in the present report. Its relations to success in agriculture, however, are such that, upon further reflection, I felt it my duty to make some effort in aid of a progress towards a reform in this regard.

The utility of farm yard manure is so universally conceded, that one would suppose a farmer would be as little likely to allow waste therein as in his crop or his money. We find, however, on many farms the most useful constituents of the stable manure either escaping into the air, or being washed off during heavy rains. I cannot but believe, if farmers would make themselves better acquainted with the *chemistry of manure*, that very many who neglect this branch of their occupation would have the extent of their loss so plainly exhibited, as very promptly to bring about an improved system.

Some years since a gentleman, now numbered with the dead, and who was alike distinguished as a statesman and a farmer, said to me: "We have various substances used for "manuring the soil, but why is it that barn-yard manure is "the only kind suited for every kind of soil and every kind "of crop?" The reply, of course, was that it was produced from the soil, and contains every constituent that enters into the composition of plants, and in such a state as to be easily assimilated by them.

Boussingault, who, besides being one of the first chemists of the day is a practical farmer on a large scale, holds the following language in his valuable work under the title of Rural Economy:

"In agricultural establishments in which the importance "of manure is duly *appreciated*, every precaution is taken, "both for its production and preservation. Any expense in-"curred in improving this vital department of the farm, is "soon repaid beyond all proportion to the outlay. The in-"dustry and intelligence possessed by the farmer may indeed "be almost judged of at a glance, by the care he bestows on "his dunghill. It is truly deplorable to witness the vast loss "and destruction of manure over a great part of the country. "The dunghill is often arranged as if it were a matter of "moment that it should be exposed to the water collected "from every roof in the vicinity, and as if the object were to "take advantage of every shower, to wash and cleanse it "from all it contains that is really valuable. The main "secret of the admirable and successful husbandry of French "Flanders may, perhaps, lie in the extreme care that is taken "to collect every thing that can contribute to the fertility of "the soil."

He also adds, that if "premiums were awarded to those "farmers who should preserve their dunghills in the most "rational and advantageous manner, they would prove of "more real service than premiums in many other and more "popular directions."

These observations were intended for France, but they are equally applicable to this country.

It is really painful to witness the want of care in this respect which so generally prevails. There are, it is true, many exceptions in those farmers who are not only fully aware of the necessity of preserving their manure, but who use every means in their power to retain all its valuable constituents.

If it were possible to find two adjacent farms of equal size, with soils exactly alike, and managed and cropped alike; each having the same number and kind of domestic animals, an experiment of the most expressive kind could be tried. We will suppose that one of these is in the possession of a person who takes "every precaution for the production and "preservation of the manure," whilst on the other the manure is well "washed and cleansed of its most valuable com-"ponents." At the end of a limited period, say 10 years, if no manure from abroad be brought to either of them, it would most likely appear that he who used well-washed manure would have "short crops," which he would perhaps charge to the seasons or insects, unless he were taught better by the operations of his more thrifty neighbor.

What I term the "chemistry of the dunghill" should be taken into consideration in any system for its preservation, but it would altogether exceed my limits to go fully into the subject at this time. Besides, farmers will find the subject is fully treated in many modern books in the hands of our farmers. In this regard Boussingault's Rural Economy and Johnston's Agricultural Chemistry should be consulted by every farmer.

Although we have so much already in print upon the subject, it may not be out of place to indicate the precautions necessary for the preservation of dunghills in the most advantageous manner. I have very often noticed the dung-yard to be a recipient not only of the water from the barn and other buildings, but also from the adjacent fields. Consequently, during wet seasons the barn-yard must either be flooded or, as I have also noticed, a ditch is opened to permit the fluid to run off, carrying with it the larger portion of the useful constituents of the manure.

On the other hand, if the manure consists principally of the dung of horses, care should be taken that it be not kept too dry, or it will, in fermenting, become extremely hot and almost entirely useless.

The barn-yard, or other receptacle for manure, ought to slope from every direction to one point, which is best to be near the centre where there should be a shallow pit or tank to receive the drippings. The opening into the tank should be covered with an iron grating or strong wooden slats, sufficiently close together to sustain the manure.

The outer edges of the yard should be sufficiently elevated to prevent the flow of waters either in or out.

A pump of simple construction, which any farmer can make, should be placed in the tank, and a piece of hose provided, or what will answer nearly as well, light wooden troughs may be made by nailing two narrow boards together and providing simple supports, so that they can be extended to all parts of the yard. By means of the pump and troughs the liquid manure may be pumped from the tank and distributed to all parts of the manure yard.

If the subsoil to some depth should be a stiff clay, there will be little or no loss by filtration, but if loose or sandy, the bottom should be well puddled, and if it can be paved at a reasonable cost it will be all the better for it.

The water from the roofs of adjacent buildings should by no means be permitted to flow into the manure yard. This can be cheaply prevented by the use of tin or wooden spouting.

If the dung of other stock besides horses be put into the same yard, it should not be thrown in at random, but each kind should be taken from the stables in dung-barrows, and be uniformly distributed over the area. Besides, in an uneven heap there will be vacancies which become mouldy, to the injury of the manure.

The more solid the mass of the manure, by being trodden down, the better, so as to prevent too rapid a fermentation and consequent waste of both ammonia and humus.

If there be cow stables or hog pens adjacent, the drainage from them should be conducted into the manure.

When, as is most usually the case, the manure is to be principally used for the wheat crop, care must be taken that it should not become too dry, and be seriously injured, during our hot, and often dry summer. This can easily be prevented (if there be water at or near the barn, as there always should be) by pumping and distributing it in the same manner as the liquid manure from the tank.

Even with all these precautions the manure will be more or less injured by exposure to the direct rays of the hot summer suns of our climate. These, of course, we can only avert by the use of sheds, except where the manure heap happens to be shaded by large trees. In Europe the cost of lumber, in most cases, forbids the use of sheds for this purpose, and in fact they are not much troubled with hot suns, except in Italy and parts of Spain and France. There are, however, numerous locations with us in which sheds might be most advantageously erected at little expense.

We learn from the direct analysis of Boussingault that manures, *properly fermented*, actually contain more nitrogen or ammonia than the dung, straw, &c., of which they are made up. This proves that there is no loss, but, on the contrary, a gain in the use of fermented manures, provided the process has been properly carried on.

It has long been known that the efficacy of farm-yard manure was increased in a remarkable degree when a very small proportion of gypsum is mixed with it, and there is abundant proof of this. It was noticed by French agriculturists in the last century, that the yield of potatoes was much increased when gypsum was added to the manure in the drills, and I have confirmed this in my own experience. More recently M. Didieux, who practices this method of using plaster, informs that his plastered manure never becomes mouldy, which is an important fact.

M. Didieux states that he uses twenty litres of calcined plasters to 2,500 kilogrammes of manure, which is equal to a fraction over one-third of a bushel to the ton of 2,240 lbs.

This small addition, he assures us, increases his crops at

least one-third both in grain and straw. I have before noticed the propriety of sprinkling plaster upon the floors of stables, and if, in addition, we add a very small proportion of copperas or sulphate of iron we get rid of the unpleasant odors, and promote the health of the animals. The proportion of copperas should not exceed one-fourth or one-fifth of gypsum, with which it may be mixed after being powdered, or it may be dissolved in water and sprinkled over the floors.

The use of peaty muck in the barn yard has been already alluded to, and is earnestly recommended to all who possess facilities for such operations.

CHAPTER XV.

Artificial Manures or Fertilizers.

The increasing demand for manures produced by the progress of agriculture during some years past, has stimulated many enterprising persons to add to the resources of the farmer in this respect. The result has been to bring into the market an infinite variety of what are termed artificial manures or fertilizers from establishments in and near the principal cities of the Atlantic States.

The numerous advertisements of the manufacturers in the agricultural and other journals, as well as their handbills and pamphlets, usually contain certificates of analysis of their samples. Many of these appear to contain phosphates and ammonia or its elements, besides other matters useful in manures.

One class of these, called "Manipulated Guanos," are said to consist of mixtures of Peruvian and Phosphatic Guanos, ground very fine, and intimately mixed together by a second grinding. The manufacturers claim, with reason, that by the perfection of their machinery, they are enabled to reduce the guanos to a finer powder, and to effect a more intimate mixture than can be done on the farm. There is no doubt that a given weight of guano in fine powder will prove more effective and be of greater value to the farmer than when coarsely ground or in lumps.

As I have before stated, there is every reason to believe, from the effects of pure Peruvian Guano, that the proportion of its ammonia is too large for its phosphate of lime. It would appear, therefore, to be advantageous and economical to use a mixture of the two, or a "Manipulated Guano."

Another class, termed Super-phosphates, consists of phosphatic guanos, (and sometimes perhaps bones,) a portion of whose phosphate of lime is said to be converted into Super-phosphate by the use of sulphuric acid. I have already expressed an opinion, (founded upon the experience of others, as well as my own,) in favor of using bones in an incipient state of putrefaction, in preference to treating them with sulphuric acid. (See chapter XI.)

The addition of this acid in *small proportions* to the phosphatic guanos, will be attended with the advantage of quickening their action. Although the name of super-phosphate of lime is applied, indicating that salt to be the predominating ingredient, but I believe that is seldom the fact, because of the cost of the acid. This is all very well, provided the farmer pays for no more super-phosphate than he receives.

A pure super-phosphate of lime consists of—

Phosphoric acid, ...	61.02
Lime, ...	23.73
Water, chemically combined, . .	15.25

From the nature of the materials used, it is not practicable to obtain the article, (*at a manure price*,) as pure as the above, which I have given for the purpose of comparison.

In addition to the two classes of fertilizers already noticed, there are numerous other compounds offered as fertilizers to the farmer. They are mostly designated by very long names, the mere enumeration of which would occupy too much space at this time.

As some of them are largely used by our farmers, I had designed to collect for analysis a number of samples of each kind, whether made amongst us or imported; but it was out of my power to carry the plan into effect, for want of a sufficient appropriation for assistance. To this time I have found it necessary to apply a large amount of my private funds to the expenses of executing the varied duties of my post.

A few could have been done, but in making selections of these, parties might chance to say that they were made through favoritism or invidiously, depending upon the result being favorable to the honesty of the maker or otherwise. These considerations, necessarily, have more force from the fact that the work, important as it is, has not been specifically made one of the duties of my office. I should, however, have felt so certain of its propriety, as not to have hesitated, if sufficient means had been supplied for the purpose.

That fertilizers can be, and are honestly prepared at prices within the reach of the farmer, I am fully satisfied; but the latter is liable to suffer from several causes. They may, in some instances, be honestly though unskillfully prepared in ignorance.

And again, they may consist principally of materials of little or no use, except to add to the weight and to the profit of the dishonest maker. It is as easy for the latter to publish flourishing certificates as for his more honest rival in the trade. Some of the makers have complained of the injury to their trade, and have expressed a desire that effective measures should be adopted to put them upon a better footing.

The mode by which I propose to effect this purpose will be stated presently, and will be applicable to manures whose properties are not apparent by simple inspection. I have already referred to the investigations which have been made of bones and guanos for farmers, and there are others in the hands of my assistant which I hope will be completed in time to be placed in the appendix.

An article has been advertised in the "Country Gentleman" paper, in praise of what is termed the "National Fertilizer," in which it is stated that "its basis is the green sand marl of New Jersey, combined with fish and pure animal bones."

As to the bones, I would think it better that the farmer should buy them pure and ground, and manipulate them in the manner I have before suggested. And if he has access to fish or flesh cheap enough for manure, he can easily compost them with earth and plaster, and the Jersey marl could be delivered to us for three dollars per ton, if we should want that article. It is apparent, where one article is very cheap, while others cost 8 or 10 times as much, that the proportions of each becomes very important, and the temptation to increase the cheaper is very strong.

I again avail myself of the results of Prof. Johnson's labors, in his recently published work, in reference to fertilizers.

It is known to many farmers that the money value of manures is fixed by that of their useful contents, a practice instituted in England a few years since. Prof. Johnson considers ammonia to be worth . . . . 14 cts. per lb.

Soluble phosphoric acid, ... $12\frac{1}{2}$ " "

Insoluble " " . . . . $4\frac{1}{2}$ " "

An equivalent amount of bone phosphate would be worth about 2.2 cts. or 2 1-5 cts. For the present we may make use of the standards of value given above, although I am satisfied that soluble phosphoric acid is rated too high. Further investigation will be required, however, to fix its real value to the Maryland farmer, which I have not had time to make.

Assuming these values for the present, we can readily determine the value of any fertilizer whose composition is known, by multiplying the percentage of each constituent by the price; and these added together will give the value of one hundred pounds, which, multiplied by the number of pounds in a ton, will give the value of one ton of the fertilizer.

This may be illustrated, as it is done by Prof. Johnson, by supposing the valuable components of a fertilizer to be as follows:—

Ammonia, 3 per ct. at 14 cts., . . .	$ .42
Soluble phosphoric acid, 11 per ct. at 12½ cts.	1.37½
Insoluble " " 10 per ct. at 4½ cts.	.45
Value of 100 lbs , ...	$ 2.24½ X 20
" one ton of 2,000 lbs., . . .	$44.90

By this means Prof. Johnson determined the value of several fertilizers.

1st. *Mapes' Super-phosphate* from Newark, N. J. In 1852 its calculated value was $44. In 1857 it had degenerated to $15, owing to the introduction of worthless matter and the total absence of soluble phosphoric acid.

Another article, called "*Mapes' Nitrogenized*," possessed a value, by calculation in 1856, of $21; and in 1857 one sample proved to be worth $14.50, and a second $12.50, so that it seems to be going down pretty fast.

The name of *De Burg's Super-phosphate*, of Williamsburg, Long Island, so familiar to farmers from advertisements, proved to be worth, in 1852, $32; in 1856, $36.25, and in 1857 it had fallen to $21.50.

Coe's Super-phosphate, from Middletown, Conn., has proven more uniform in composition, as shown by seven analysis between 1854 and 1857, its value being as follows:—$33.75, $36. 25, $33, $41, $33, $35 and $33.25.

Prof. Johnson calculated the value of Rhodes' Super-phosphate of lime, (a Baltimore article,) from three analysis, to be $32.25, and his results, he remarks, do not seriously differ from those of Dr. Higgins and Bickell.

Jourdan's Super-phosphate.—Since this chapter was placed in the hands of the printer, Dr. Piggot has reported to me the results of analysis of two samples of an article under the above name. They were furnished by Maj. Edward Wilkens, of Kent county.

The first was purchased in 1858, and was used with good effect by many farmers in that county. The second was purchased in 1859. Their composition is as follows:

	1858	1859
Gypsum or plaster of Paris, . . .	25.30	39.31
Soluble phosphate of lime, . . .	2.53	2.95
Free phosphoric acid, ...	6.86	4.47
Lime, otherwise combined, . . .	2.07	

	1858	1859
Phosphoric acid, combined with lime and magnesia, ...	——	—— 2 23
Sand, ...	11.04	14.30
Animal charcoal and organic matter, (containing some ammonia,) .	22.30	12.32
Magnesia, iron, water, &c., not determined, ...	4.66	13.95

The useful matters may be summed up as follows, and I have also calculated their money value in the manner before stated. (See page —.)

That of 1858:

	Per ct.	Price.	Am't.
Gypsum, ...	25.30	⅓ ct.	$ .08½
Phosphoric acid, insoluble, .	11.65	4½ ct.	.52½
" " soluble, .	8.40	12½ ct.	1.05
Value of 100 lbs. of the fertilizer, . .			$1.66

That of 1859:

	Per ct.	Price.	Am't.
Gypsum, ...	39.31	⅓ ct.	$ .13
Phosphoric acid, insoluble, .	7.06	4½ ct.	.31¾
" " soluble, .	6.27	12½ ct.	.78¼
Value of 100 lbs. of the fertilizer, . .			$1.23

The value of 100 lbs. being multiplied by 20, gives the value of a ton of each.

Thus, that of 1858 is worth	$33.20
" 1859 "	24.60
Difference against the latter, ...	$ 9.60

The proportion of ammonia was too small in either to be worthy of notice.

Comment is unnecessary. I have given the chemical constitution and its money value, so that the farmer may really know what he is buying.

The result of all this shows a *great falling off in the value* of three manures, which have been much used in Maryland, whilst in two others, (Coe & Rhodes,) the quality has been generally maintained.

What quantities of these inferior articles have been sold to our farmers because of their original reputation, cannot be ascertained, but it would seem that means should be taken to arrest such frauds. It is felony to obtain money or goods under false pretences, and people are punished criminally for such acts. Is it not equally criminal in morals, if not in law, to publish certificates of the existence of certain proportions of valuable matter in a manure, and yet sell a material containing perhaps one-third or one-half the amounts stated in such certificates?

After consultation with many farmers and planters, and seriously reflecting upon this subject, I am fully satisfied that if a proper sum be allowed me for such assistance as will permit a comprehensive system of analysis to be executed, the evil will be very soon corrected.

The conscientious maker or dealer will, of course, furnish fair samples; but as there might be some who would act otherwise, I would propose to take such means as would insure samples of the articles *actually received* by the farmer. Such a number of each kind used in this State should be analyzed from time to time as will keep the public informed of their composition and value. The law might require these to be reported monthly or quarterly to the Governor or other officer, and published in one or more papers in each county, as in the case of the laws.

The effect of these measures would not fail to afford ample protection to the farmer against both fraud and ignorance, and whilst benefiting the honest dealers, would very soon drive all others out of the trade.

I fully accord with Prof. Johnson, also, in the opinion that in estimating the money value to the farmer of these *costly* manures, we may disregard all their constituents, except ammonia and phosphoric acid. If the proportions of these be correctly given, any farmer can readily calculate the real value of each, and determine which it is his interest to purchase.

A sample of a fertilizer was forwarded to me by Charles S. Contee, Esq., under the name of "*Grass Manure.*" It was analyzed with the following results:

Common salt, (chloride of sodium,)	44.03
Phosphate of lime,	27.63
Phosphate of magnesia,	3.53
Carbonate of lime,	10.51
Carbonate of soda,	.97
Organic matter,	4.44
Water,	7.92

The above results indicate that this "grass manure" consists of a mixture in nearly equal proportions of good Mexican guano and common salt, and is doubtless useful as a manure. Its money value can therefore be easily determined.

CHAP. XV.

CHAPTER XVI.

Observations upon the present state of Agriculture in Maryland, with suggestions for its improvement.

Upwards of two hundred and twenty-five years have elapsed since Leonard Calvert and his small band of pilgrims effected a settlement on St. Mary's river. They honestly purchased lands from the owners, including the clearings upon which the Red Man had raised his tobacco and corn. Then was commenced a most exhausting system of agriculture, which was continued by them and their successors during a period of little short of two hundred years. The result of this old system of almost incessant cropping without manure, reduced the productiveness of the soil to such an extent that at length the crops would in many places scarcely bring the cost of raising them.

Such was the condition of much of the soil in that region when I visited it, about 26 years since; but I had an opportunity in the early part of the last summer to see the same lands bearing heavy crops of wheat, yielding, perhaps, four times the amount per acre as when I first visited that part of St. Mary's county.

Dr. Broom, an extensive as well as successful farmer, (who now owns and farms the land upon which the first settlement was made,) stated that this remarkable improvement in the soil was mainly caused by the use of lime, and, I may add, by intelligent management.

This locality is mentioned, because it contains the first land cultivated in the State by the European settlers. The settling and culture of the other tide-water counties soon followed, and with a similar impoverishment of most of the soil. By the improvements of modern agriculture, so generally adopted by our agriculturalists within the last 30 to 40 years, the improvement of the soil has progressed in an increasing ratio.

If the United States census statistics were correctly given, we should find when the returns of 1860 shall be published, that a large increase of products will have been realized during the preceding ten years throughout nearly the whole State.

It may be useful for us to inquire into the causes of this

improving condition in the *great interest* of Maryland. These will be found probably as follows:

1. An increasing desire for *correct* knowledge in reference to the art of culture, which has resulted in making the intelligent farmer better acquainted with the principles involved in the art or science of culture, and from this most of the other causes flow.

2. The extensive application of lime and marl, which developed those constituents of plants contained in the soil in an insoluble or inert condition, and rendered them fit to be taken up by the plants.

3. The use of guano, bones, ashes and other fertilizers which supply ammonia and phosphoric acid, which (not originally abundant) were taken out by the crops to such an extent as seriously to exhaust our soils in reference to these constituents.

4. The introduction of machines for raking, reaping and threshing, and other purposes, by which much more land can be tilled by a given amount of manual labor than formerly.

5. Another and very efficient cause is an increased interest in the closer personal superintendence of the daily routine of the farm or plantation.

6. The introduction and extensive use of clover.

It is by no means to be supposed that the march of improvement has ceased; on the contrary, the advancement already made only serves to stimulate to further progress. We know that 50 to 60 bus. of wheat have often been raised per acre in England, and we are told that even the rate of 90 bus. has been reached on small lots. It *is* therefore practicable to produce these large yields, and we must push on with our investigations until we become fully acquainted with all the circumstances relating to this interesting subject. It is only by carefully studying the art of culture in connection with its principles, as derived from other sciences, that we can ever hope to attain maximum results from the soil.

That the practice of agriculture is still too empyrical, must be evident to those who examine into its present state.

One farmer, for instance, will insist that by the use of lime only his land will always continue fertile. His assigned reason is, that he knows it to be so, because he has learned the fact by his own practical experience. He might with as much propriety assert that if he had a quart of sand he could continue taking out a grain a minute forever.

Another, who, after using lime for a long period, finds it less effective than formerly and resorts to guano, and after two or three successive good results from Peruvian guano, adopts as his theory that guano will always keep his land productive.

I might relate numerous analogous views equally erroneous, that have come under my notice, each derived from *practical experience ;* but it is unnecessary, as there is scarcely an intelligent farmer that cannot find such examples of false inferences.

I have stated why it is that the manure from the barnyard is almost universally applicable to soils, which have not already an abundance of every kind of plant food. I may add, that it is the only article of its class.

If we have a soil deficient in all, or most of the constituents of plants, and have not the stable manure, its place may be, in a great measure, supplied by the use of bones and wood ashes. But we should not depend solely on either these or the dung. It is true that in both cases we supply every thing the crops require until one or more of their elements shall be exhaused, but we must remember that they take little or no part in developing inert matters previously existing in the soil. For this purpose we must apply lime, unless we may prefer to incur the expense of supplying from abroad every element which the crop must take out of the soil.

I have already referred to the conclusions that most chemists have arrived at, in reference to the analysis of soils for practical applications; but I find I have omitted to notice a very remarkable case of a soil from the Island of Cuba, containing 90 *per ct. of oxide of iron.* The sample was analyzed by Prof. Hayes, of Boston, and was from a soil producing fine crops of tobacco. The result seemed so strange, that I made inquiry of the Doctor in reference to the truth of the published statement; which he confirmed. Now, if any chemist were asked to name an article containing 90 per ct. of oxide of iron, would he call it a soil? Would he not call it a very rich iron ore?

This case, with many others that might be cited, confirm the views before expressed that there is something more to be known before we can rely wholly upon an analysis of soils for practical applications, except in the special cases alluded to in a former chapter.

In reviewing the present state of agriculture of Maryland, many considerations of great interest must come under our notice. Some of these which have been touched upon in the preceding pages will be again referred to in connection with the rotation of crops and other subjects.

Nature plainly teaches us the necessity of her practices in reference to what we call the "rotation of crops." Every observing farmer must have noticed that, with the exception of a few plants whose roots extend deep into the ground, there is a constant succession or interchange of place among plants on uncultivated lands. Take for example a field set in timothy. If the soil be very rich, and the crop has been properly put in, we shall have a most luxuriant growth, consisting almost wholly of the timothy, during several years, after which, other grasses or plants appear, and if the field be pastured pretty freely the timothy will soon be entirely eradicated. In our climate and soils, if the pasturing be continued, the whole field, unless it be very poor, will be mainly occupied with white clover and green grass, (poa pratensis,) improperly called by some "blue grass," which is a different plant. When we mark the boundaries of some adjacent little patches of each of the two plants by placing sticks around them, we find in the course of a year or two a complete interchange of place between the grass and the clover. If the land be limed every eight or ten years and top dressed heavily every four years, with stable manure or city street dirt, it will produce rich pasture for many years. There are fields of this kind in the south-eastern parts of Pennsylvania, from which for 20 years or more large crops of hay have been annually gathered in June, and which afford afterwards rich pasturage until covered by snow during the winter.

If instead in a rich soil, the timothy had been sown in one deficient in one or more of the constituents required by this grass, other plants or weeds which can find all their elements in such a soil, will soon spring up among the timothy. This grass sometimes will share the field with the intruders for a time, but as the soil is more and more exhausted, the timothy will finally disappear. And although it would be labor lost to sow timothy in the field again without supplying appropriate manures, yet in many cases some other crops will succeed tolerably well.

Other illustrations of nature's principles of rotation may be drawn from forests. In most of these we find a great variety of trees and shrubs intermixed ; some of which derive their mineral constituents from near the surface, whilst the roots of others penetrate frequently many feet beneath. The necessary mineral matters are thus taken up by each kind of tree or shrub, and portions of them are annually distributed as a top dressing, in the leaves, flowers, fruit, &c. The trees and shrubs themselves (if left to nature) die after completing their growth ; and in their decay leave upon the surface their stores of plant-food for the use of their successors.

If only a single species, or at most one or two, had existed in a forest, we may be certain that they would, after one or two generations of the trees, dwindle and finally give place to other kinds, whose mineral constituents differ in their proportions.

The only apparent exception to this law of nature, is in the growth of pine forests, but this is more apparent than real, and is owing to the smaller proportion of mineral constituents required by the pine, which can, therefore, forage longer in one place than the deciduous trees.

We find, however, in many situations, if the ground be left undisturbed after removing a forest of pines, that trees of other kinds will succeed. There are soils so unsuited to any other than pine trees, that they will again spring up from the scattered seeds; but so far as we know at present, this second growth is never thrifty.

The cutting of wood for the use of iron works in Maryland commenced nearly 150 years ago, and in many cases trees were permitted to grow up again.

It was the practice of the iron masters to harvest a crop of wood once in twenty years. We are without records of the characters of each crop, but the traditionary testimony is to the effect that the kinds of trees which predominated during one of these periods materially differed from their immediate predecessors.

The preceding facts, which have been given somewhat in detail, show that a succession of plants in the same land is one of the inflexible laws of nature. It is true that there are cases in which certain kinds of crop had been raised for many years in soils abounding in all the constituents of plants, but these are very rare exceptions. Every farmer knows that in attempting to raise wheat on the same field from which he has taken a crop of this grain the preceding year, that it will not turn out well unless the soil be very liberally supplied with the proper kind of manure. And if he persists he will have less and less every year.

Severe cropping with grain attracted the attention of Charlemagne more than 1,000 years ago, who by law required that no land should be cultivated more than two years out of three. During the third year there was of course a crop of weeds, the decay of which assisted the growth of the succeeding crops, and time was given for mineral matters to become soluble and available to plants. Maryland planters and farmers were soon obliged to adopt the same rule, leaving each field for one year out of two or three to a growth of weeds, formerly called pasture. Upon the introduction of clover, which was sown upon the wheat land, we acquired a valuable forage plant in the place of the weeds. This formed an era in our agricultural history which requires at least a passing notice.

The principal money crops, as they are termed, which are

cultivated in Maryland, happen to consist of plants whose roots penetrate very little beneath the surface of the ground. Clover has long tap roots, penetrating beyond the depth reached by the plow, even into the sub-soil, by which means it is enabled to appropriate matter altogether inaccessible to grasses, grain and tobacco. Under our old system of agriculture, including frequent working of the soil, large portions of the mineral constituents of plants were rendered soluble, and in part carried by rains into the sub-soil.

The deeper plowing of later times has brought much of these within the reach of our grain and other crops, but clover is an important agent in effecting this purpose, because its roots go much deeper than the plow. In addition to the mineral food which clover thus brings to and near the surface, it supplies in its decay a large amount of humus, which, as has been already stated, is required for the production of heavy yields of grain, tobacco and other money crops.

The introduction of gypsum soon succeeded that of clover, and was found very efficient in promoting its growth. It was next supposed that with the aid of clover and plaster, no other means need thenceforth be resorted to for sustaining the permanent fertility of the soil. But it was found afterwards that if clover be too frequently repeated in the same soil it did not flourish so well at first, and the term *clover sick* was applied to such soils; although it was difficult to imagine how a soil can get sick.

The truth is that a too frequent production of clover must tend to exhaust the soil within reach of its roots, of such of its mineral matters as are in a condition to be available for the use of the plant. Upon the decay of the clover its remains are used up by subsequent crops and are in part removed, so that the reduced amount of clover, as well as of other crops, is owing to a deficient supply in the soil of matters essential to their growth.

This state of things may be remedied in part, at least, by supplying the mineral deficiencies, but the most effectual and economical method is to alter the rotation, if it can be done consistently with the present profitable working of the land.

In England the introduction of turnips into the rotations upon light soils has had a most beneficial effect, for besides the profit in feeding these roots, much more grain is now raised on the same farm than before the introduction of turnips as a field crop. Our climate is not adapted to this plant, and although there are roots, such as parsnips, and carrots, &c., that we can raise without difficulty, yet they are not likely to make part of our field crops, unless the raising and fattening of stock should be engrafted upon our system.

Experience has satisfied English farmers that one grain crop should never immediately succeed another. The intercalation of a green crop (consisting of either grass, roots, peas, beans or clover, &c.,) between crops of grain, is considered absolutely necessary to preserve the soil from exhaustion. How long our newer soils will bear the system of farming with the rotations now prevalent in most of our counties, it is impossible to predict. We are unquestionably lessening the amount of plant food in the soil faster than its materials are being separated from their combinations, and made available to our crops.

It behooves farmers to turn their attention seriously to this subject so vitally important to our state; and to endeavor to change our present system for one better calculated to sustain the fertility of the soil.

In some of our counties, we find the following rotations:

	A.	B.
First year,	wheat.	corn or tobacco.
Second "	corn or tobacco.	wheat.
Third "	wheat or oats.	clover.
Fourth "	clover.	corn or tobacco.

In each of these we find grain crops succeeding one another, except on plantations where part of the fallow crop is tobacco, which, although less exhausting than wheat, is repeated too frequently for most soils. In the first case (a) we have three successive exhausting crops to one crop of clover—or clover once in four years. In the second, (b) clover comes in every third year.

In some of the upper counties the rotations are as follows:

	C.	D.
First year,	corn.	corn.
Second "	oats.	oats.
Third "	clover.	wheat.
Fourth "	"	clover.
Fifth "	wheat.	"
Sixth "	timothy and clover.	timothy and clover.

The timothy is permitted to remain two years and upwards, according to circumstances.

The first, (c) is calculated to sustain the soil for a very long period, provided the clover, timothy and straw, be consumed on the farm. The second, (d) includes three successive grain crops, and, except on rich soil, requires an abundance of manure to secure the wheat crop, which is the third in the succession.

There is one good feature in both, which is, that the soil is copiously supplied with humus by the clover and the grass, and these being consumed on the farm by the stock, increase the amount of manure, and consequently the crops of grain. A practice prevailed at one time in parts of Harford and perhaps other counties, of sowing clover seed in the corn field at the time of the last working. In the spring and summer of the year following, the clover was eaten off by cattle for the butcher, and the ground sowed in wheat, which, in such cases did well, when the crop of clover was good.

This system would be a good one, but for the hot and dry weather we often have about the time the clover must be sown. Some kinds of peas are used for similar purposes, and it is more than likely, that they or some other plant might be sown in the cornfield with advantage, and after the corn is secured, be either plowed in or eaten off by cattle.

Whether this be done or not, the three or four field rotations should by all means be amended by the intercalation of grass for hay or pasture, in which it should remain *not less* than two years. With some such improved rotation, we would, doubtless, raise as much grain and tobacco in the aggregate as we now do, besides supplying a large amount of meat to add to the farmer's income. In former days, beef, pork and mutton, continued at very low prices for many years, owing to the supply from the West, but those who have watched the progress of affairs, will readily conclude that the days of cheap meats, especially beef and mutton have passed away, never to return. Farmers in some of the counties are well satisfied of this, and are shaping their course accordingly. They find also, that by the introduction of forage plants more frequently, they greatly augment their barnyard manure. It is certain by this course they will, in a great degree, arrest the exhaustion of their soil, increase the amount of their money crops, and be enabled to add to their receipts by the sale of cattle, sheep and hogs.

It may seem out of place for me, to recommend particular crops, or system of farming to my agricultural friends, but in addition to those already referred to, there are some considerations connected with the culture of Irish potatoes, which I deem it proper to submit.

It is well known, that, of the large quantity of Irish potatoes consumed in the city of Baltimore, and its vicinity, and supplied to shipping, a very considerable proportion is imported from other states, and Nova Scotia. It would seem that, as our climate, and much of our soil is well adapted to raising this crop, we ought surely to supply our wants and even have a surplus for export; this, doubtless, would be the case, if the matter were properly examined into by our farmers.

In the light soils of the green sand marl districts of New Jersey, both Irish and sweet potatoes have for a long period been extensively cultivated for the supply of New York, Philadelphia, and other places, and the culture of them is rapidly increasing.

The quality of the Jersey Irish potato for table use is so excellent, that they uniformly bring higher prices in the city markets, (especially the peach-blow variety), than those from other districts.

More than a year ago, there was a discussion in some of the agricultural papers, in reference to the cause of the good quality of the Jersey Irish potatoes. Among others, there was an article published in the *Country Gentleman* of January, 1859, which throws some light upon the matter.

After giving statistics, showing the annual product of potatoes to have been nearly quadrupled in four counties, between 1840 and 1850, the writer adds the following:

"I suppose that the preference given to Jersey potatoes, is "owing to the fact, that green sand is almost the only manure "used in raising them. It is a common opinion in New Jersey, "that potatoes raised from marl are much better for the table, "than those from heating manures." He adds, that the "Peach "Blow, a *late variety*, will be found to succeed first rate far-"ther south."

Whether the best mode of applying the marl be in the drill, or broadcast, is not stated; but further information will be obtained upon this subject.

It is to be hoped, that those who have access to the marls of our State, will make trials of it for this special purpose. It is more than probable, that the green sand containing shells which I have described, will answer better than the Jersey green sand, in which there is a deficiency of lime and phosphoric acid.

Modes of Applying Manure.

In supplying the soil with such manure as may be required to increase its productiveness, we should be careful to apply it in such manner as to avoid a waste of any of its useful constituents so as to obtain from them the greatest maximum effect. In the investigation of this subject we must have reference to the characters of each kind of manure. I shall notice in this connection some of the most important manures now in use.

1.—*Lime.*

If we obtain lime freshly calcined and in lumps, it should be thoroughly slaked, which, if the lime be pure, will reduce nearly the whole of it to a powder. It is then in a state of what is termed hydrate of lime and is soluble in about 600 times its weight of water. When by a longer exposure it is converted into carbonate of lime, which is soluble in 10,000 times its weight of pure water; but much more soluble in water containing carbonic acid, as is always the case with rain water. These properties of lime indicate, that if plowed deeply into the soil, as was formerly the practice with many persons, much of it is entirely lost, because the percolation of water through the soil will dissolve and take it out of reach of the roots of plants.

When uniformly spread upon the surface the action of the frost tends to crumble down the lumps and grains which may have been imperfectly slaked, whilst the rain water gradually dissolves and distributes the lime throughout the soil.

It is manifest, therefore, that it should *on no account be plowed in*, but kept as near the surface as possible. If it be necessary to apply it just before a crop is sown or planted, the ground should be previously plowed, and *horrowed also*, to prevent the lime getting down between the furrows. Experience has abundantly shown that the most effective applications of lime are those in which it is spread uniformly over a field in grass at least two years before it is to be plowed. Where grasses are not cultivated the lime may be applied to the clover ground in the autumn of the season after it is sown. Even if this limed clover sod be plowed for wheat in the year following, the good effects of the lime will be of a decided character, but still more so if the plowing be deferred to the second year, for either fall or spring crops.

If land capable of bearing a fair growth of clover, be treated in this manner, and the clover be not cut but pastured for two seasons, the productiveness of the soil will be found much increased. Under the old system of spreading lime, without paying proper attention to slaking, and immediately putting it in with the plow as far out of reach as possible, it was quite common to apply 100 bus. to the acre (and sometimes even more) of the fresh pure lime of Baltimore county. This, when slaked, will increase in bulk to from 200 to 250 bushels. At the present time the conclusion of most farmers seems to be that, if properly applied, one-fourth to one-half the quantities at a time answers fully as good a purpose. It is my belief that under the *burying* system the benefits of at least one-half the lime was utterly lost, and of course the farmer was a loser to the amount of one-half the cost he paid for it.

2.—*Marl.*

The same causes which indicate that the soil upon which lime has been spread, should not be plowed within less than 18 months or 2 years, are equally applicable to marl. If the arrangements of the rotation will not permit the application to be made to the clover or grass sod, we may harrow it as in the case of lime; but in using marl it is very important that it should be exposed one, or better two years on the surface. By this means the harder shells will be more completely disintegrated and become more intimately mixed with the soil.

3.—*Bones.*

These should be carefully distributed either by hand or a suitable machine, upon a surface which has been plowed and harrowed, and then harrowed in so as to be well mixed with the soil. It is necessary that they be covered in the soil to the depth of not exceeding two, or at most three inches, so that they may have sufficient moisture to continue the production of ammonia and the solution of the phosphate of lime. If left on the surface the chemical changes are arrested whenever the bones become dry; and when renewed by moisture, most of the ammonia escapes into the air, except during rains, when it is carried into the soil. If they be *buried* by the plow, their immediate effect is seriously lessened, because the changes go on very slowly, and the products are liable to be carried out of reach of the roots of plants almost as fast as they are developed.

4.—*Guano.*

It was a common practice to plow Peruvian guano deeply into the ground, by which there was an inevitable loss because of the solubility of ammonia and other soluble matters. This loss was not so apparent when 400 to 500 lbs. were applied to the acre, but the case is different when 150 to 200 lbs. are plowed down. Probably the practice of plowing was in order to avoid the loss of ammonia which happens if it be left on the surface. But in avoiding this it is not necessary that we should lose by going to the other extreme. Good results have followed drilling in *small* doses of Peruvian guano with wheat and other small grain. This is a good practice for a costly and evanescent manure, but care should be taken to avoid an excessive dose, which will prevent the germination of the grain or kill the young plant.

Less than three bushels of gypsum, if thoroughly mixed with a ton of Peruvian or other ammoniated guano, will in a great measure prevent the ammonia from being volatilized.

The phosphatic guanoes have nothing to lose by evaporation, and therefore may be left on the surface; but it is better that they be harrowed in also.

Common salt, if not mixed with bones or other manure, may be sown on the surface. Being very soluble, it is soon carried into the soil. The same may be said of all the alcaline salts, or other very soluble manures.

5.—*Stable Manures.*

The objection to burying stable manures are similar to those which forbid plowing in bones. If to be applied just previous to putting in a crop, the stable manure should be spread, after the ground shall have been plowed. It may then be worked into the soil with a heavy harrow, a drag, or by means of a *very shallow* plowing. It may be urged that the expense of hauling the manure over the plowed field increases the cost and labor; but this is a small matter compared with the loss sustained by burying the manure too deep in the soil.

When we desire to possess a rich and *permanent* pasture field, we can in no way more effectually and economically produce it than by a liberal top dressing with stable manure. This is the practice in the south-eastern counties of Pennsylvania.

Conclusion.

In order to make the present report more immediately useful to the farmer, I have avoided, as far as possible, scientific discussions and descriptions. These will more properly be brought into view in a final report.

It was intended to have treated in a much more comprehensive manner most of the subjects, and to have introduced many other matters that I believe interesting and useful to the farmer and planter. It appeared, however, impracticable to do so without materially increasing the size of the report. It became, therefore, necessary to treat many subjects in the briefest manner and to omit others, in order to compress the report within what might be considered proper limits.

The field work was seriously retarded, as I have before stated, by the inclement weather of last year, so that in order to complete my programme as far as possible, I continued in the field to a late period in the autumn. Before proceeding to write out the report, it became necessary to construct the map of "geological illustrations," on account of the references to be made in the report. This was a work of much labor, occupying my whole time for more than a month, when it was put in the hands of the engravers. Being the first work of the kind which had been undertaken in this State, it was also necessary that I should look after it whilst the lithographic stones were being prepared for the printing.

The construction of the large map also required much attention in collecting information and in having it used so as to insure accuracy as far as practicable.

The necessary attention to these maps delayed the report, so that I was wholly unable to present it until the 10th of January.

I much regret that the haste with which I have been compelled to close did not permit a final reading, with a view to making verbal corrections, which doubtless will be required. Such oversights will be readily noticed by the reader, and will not impair the meaning of the sentences.

The subjects, it will be observed are treated with especial

reference to agriculture, but I have abundant evidence that our citizens expect still more of me. They require information to aid them in making available our varied and abundant resources applicable to other branches of industry, which I have in all cases cheerfully given.

I am fully convinced that a complete description of all the mineral resources of our State would tend greatly to promote the public interest. It is therefore a source of great regret that I have neither time nor space to add such an account to this report. Unwilling, however, to neglect these important sources of wealth, I shall give in the appendix a brief sketch of some of the most prominent.

I shall also add some views in reference to Artesian wells, and of their adaptation to different parts of our State.

It is my intention on a future occasion to present a full exposition of all our resources, which I think will show that there is not a territory on the face of the globe, of equal extent, whose sources of wealth are superior to those of our own Maryland.

The geological constitution of our State is such as to give us every formation that exists in any other State. By means of our noble bay and tide water rivers, the cheapest means for transportation are supplied to more than one-half the State, whilst the best means that art can give have been and continue to be applied to the remainder. Upon the completion of the new lines of railroad now contemplated, every farm in the State will be within a few miles either of a railroad or of navigation. These, taken in connection with our agricultural and other industrial resources, when fully developed, cannot but produce a degree of prosperity no where to be excelled.

I cannot close this report without expressing my warmest acknowledgments for the kindness I have every where experienced during my travels in the counties, as well as my gratification at the interest expressed in my investigations. This interest was constantly indicated in offers of every facility that might be required. By the Baltimore & Ohio Railway Company this interest was manifested by giving the privilege of free travel over that important work.

APPENDIX.

MINERAL RESOURCES OF MARYLAND.

Such of the preceding chapters as are descriptive of the geology and the mineral matters of the State, have reference exclusively to their direct application to agricultural pursuits.

We have now to consider other mineral resources, which, if less directly related to agriculture, constitute important adjuncts by furnishing materials for industrial operations, which add to the amount of our productive industry, and consequently to the wealth of the State.

There are probably few persons aware of the extent and variety of the mineral resources of Maryland, and it surprises those who have examined into the matter, to find many of them so much neglected.

Citizens throughout the counties certainly expect my aid in this matter, if I am to judge by their numerous inquiries about substances existing, or supposed to exist on their lands.

My attention is often called to such matters while executing the field work of my duties; samples are brought to me, and many inquiries are made in person and by letter. I have taken great pleasure in giving the necessary information in such cases so far as my long experience in these matters has enabled me to do.

Replies have been made to all inquiries except those recently received, and which will be attended to as early as possible. The preparation of the report and maps, have wholly occupied my time for more than two months past.

In order to aid in dissemenating information upon these subjects, I propose to give a very brief account of those mineral substances of our State at present known, which possess an industrial value, and which have not been described in the preceding chapters.

I. MARBLE.

Every kind of limestone which admits of being smoothly dressed or polished, is called *marble.*

The limestones suited for producing lime, having been already described, I shall in this place notice only those applicable to architectural and other uses.

We have in Maryland a much greater variety of marbles than is usually supposed, but many of them have not yet been sufficiently explored, and with the exception of the white marbles of Baltimore county, little effort has been made to bring them to the notice of those likely to develop and give them a productive value.

The marbles of Baltimore county constitute part of the metamorphic limestone (No. 11) on the map. Those used for marble may be divided into three varieties.

The first is the fine grained white, or nearly white marble, such as that used in the construction of the Washington monument and for buildings, and other architectural uses in Baltimore.

The best of this kind is in the vicinity of Texas and Cockeysville, lying in nearly horizontal strata, and which can be taken out in large solid masses by the use of what are termed *feathers and wedges* by the operators.

This is largely used in Baltimore, and also in Washington, both for private and public buildings, and most of it consists of a strong and durable stone. The colossal statue on the Washington monument in Baltimore, consists of marble of this kind from near Texas.

There is also in that region some excellent marble in the limestones, whose strata have more or less dip, or inclination from the horizon; but there are intercalations of other layers of less purity and beauty, which must be removed, and the dip of the strata soon brings the stone down to the *water level*, which considerably increases the cost of quarrying the stone from the necessity of pumping out the water.

I need not give a minute description of this stone, because there are few or none of my readers but what have often seen it. It is a beautiful stone for all situations wherein a white marble is required.

The second consists of limestone more or less colored, generally of some shade of gray or bluish gray. Some varieties are liable to be in part stained yellow, owing to the oxydation of the minute grains of iron pyrites disseminated through them. I have noticed this defect also in some of the white marble, and which, of course, impairs its beauty. These kinds may be avoided by a careful inspection, by which the grains of pyrites may be seen.

The third variety is that called *alum limestone*, which has already been noticed in chapter VIII, as the material from which the nearly pure lime from Texas is produced. It came into use as a marble a few years since, and was applied to the construction of the patent office and the monument in Washington. It is very white, and whilst the large crystalline grains are supposed to improve its appearance, they increase the cost of dressing it. They require the use of large files in dressing up the edges and corners of the stone, in order to prevent the grains from being knocked out in chiseling near the edges. Some of these alum marbles have been proven by experiments at Washington to be very strong; but as others are prone to disintegrate, great care should be taken in selecting them.

The Northern Central road passes through this formation, and can afford facilities for cheap transportation both North and South.

I am not aware of the existence of good marble in the limestones of this range in Harford and Howard counties.

The next range of limestone useful as marble, are on the western flank of Parr's ridge, extending southwestward from a little northwest of Manchester, in Carroll county, passing near and west of Westminster, and extending to the vicinity of New Market in Frederick county. They are usually stratified and consist of very small crystalline grains, and are generally white or of some light shade of blue. We find it, however, towards the southern limits of this range more variegated, with shades of red less pure, and the stratification more disturbed.

The different layers of this vary considerably, and even in the same quarry there are layers of both pure white and light blue, and sometimes variegated with light and dark shades of red. They take a fine polish, and are free from the grains or masses of quartz and other minerals which sometimes exist in the older limestones we have already noticed.

The cost of transportation has hitherto prevented these marbles from being transported to distant markets and they have, therefore, but a local use. The Western Maryland Railroad, it is hoped, will soon be extended to this fine region and enable those owning this material (equally valuable for marble or lime) to extend its uses.

The quarries yet opened have penetrated but a small depth; but we may expect that the demand for stone and lime, which cheap transportation upon the railroad must produce, will soon increase their depth. The effect of this will be to bring to light the marble, less acted upon by the weather, at much less cost than when large quantities of stone have to be quarried and thrown aside in order to get *unaltered* blocks of marble of large size.

Another important lime-stone, ranges in Frederick county from near Woodsboro' to the Potomac, being the most eastern range of that numbered (10) in the table in chapter III. It lies principally on the west side of the Monocacy, but that river cuts through its northern limits as shown on the map.

It is an unusually pure lime-stone, and when in former years the manufacture of Chloride of Lime was carried on in Baltimore, the best lime that could be procured for that purpose was from this formation.

It varies in color from blue to dove color, and having a very fine grain and also a close, compact texture it takes a beautiful polish and may be ranked among the finest marbles of its class in the State. The beautiful shades in the lighter or dove-colored varieties are similar to the celebrated Italian marbles of that class, and we should no longer have occasion to import marbles of that kind if those of Frederick county were properly developed.

I have before me a specimen from the quarries of William Richardson, near Buckeystown, which is more beautiful than any of the Italian of that kind I have ever seen. Samples of it were exhibited at the State Agricultural Fair, at Frederick, in October last.

The blue shades in this marble are of vegetable origin, and therefore permanent, whilst those colors resulting from the presence of metallic oxydes are liable to change upon exposure out of doors.

In the vicinity of Mount St. Mary's College, at the foot of the Catoctin Mountain, there is a similar lime-stone, and also another resembling the beautiful *verde antique* marble. There is a marble variegated with light and dark shades of red near the line of the Turnpike from Frederick to Hagerstown, about three miles west of Frederick. I have not yet had an opportunity to make a full examination of these three last, and must therefore defer further notice of them for the present. I must add, however, that the Tennessee variegated reddish marbles which are exported to the Atlantic cities and so largely used in the Capitol and other public buildings in Washington, are far from being equal in beauty to that of Frederick county.

Another marble exists in very large quantity on the western side of the Monocacy Valley and was noticed with the other lime-stones. It is widely known, because of its having been used for the large columns in the Old Representative Hall at Washington. It has received various names, including that of Calico Rock, Potomac Marble, &c.

In its characters it is something between a conglomerate and a breccia and has been much admired. There are difficulties in dressing it, which increase materially its cost, and it is only useful in columns or other large and thick pieces.

Its structure was described in chapter III, and I may now add that some of its pebbles or masses are not strongly cemented together and are liable to be loosened whilst being dressed.

There is no lime-stone or marble in Catoctin mountain, but its porphyries and amygdaloids are deserving of the attention that I propose hereafter to bestow upon them. Some of them will receive a beautiful polish but their hardness renders the process expensive. This can, however, be overcome by appropriate machining.

Crossing the North mountain we reach the great lime stone valley in which Hagerstown lies, and which has already been described.

Along the eastern edge of this we find a beautiful White Marble which ranges southward from near Boonsboro', and resembles the first or most eastern range in Frederick county. Some of it has a pure white color and a texture equally fine with that of best Italian statuary marble. It is, in my opinion, only necessary to penetrate to a proper depth in order to procure marble for statuary and architectural purposes that will give us no reason for importing the material from Tuscany.

The main body of the lime-stone of this valley has a fine grain and is free from grains of quartz and other impurities that interfere with the polish. Much of it, therefore, takes a fine polish without difficulty. The public taste, at this time, does not bring those blue lime-stones prominently into use. Some of them, however, have a very dark color and are required when Black marbles are wanted.

Westward of the North mountain we have several ranges of lime-s'one in Washington and Alleghany counties which are numbered 15 in the table in chapter III.

The different beds of these vary considerably in composition and characters. The lower strata contain the material for hydraulic lime, which will be noticed in another page.

Some of the upper layers abound with casts of fossil shells and corals. The mass of the stone is black or dark blue, whilst the lighter color of the fossils gives the marble, when polished, a beautiful appearance.

The bed of limestone in formation, No. 18, immediately under the coal formation, is not likely to be available as a marble. Those intercalated among the coal strata, are in beds too thin to give promise of being profitably worked as marble, although I have seen highly polished mantle pieces which were admired by those who prefer black marble for such uses.

In order to assist in bringing the many varieties of Maryland marbles to the notice of those disposed to operate in this branch of industry, I propose to make a collection which shall embrace each kind in every locality, in order that dealers in marble may know where the can be had.

For this purpose I beg leave to suggest to all parties owning marble, that if they will forward me suitable samples of each variety they possess, I will take the proper means to exhibit and bring them into notice.

The samples should be one inch thick and six inches square, polished on one side and dressed smoothly on the other sides, so that they may be properly placed in a case.

III. HYDRAULIC CEMENT.

As before stated, the limestone, No. 15, has among its lower beds some layers of great value, because when properly calcined and ground, they produce hydraulic lime, or that which will harden under water or in damp places.

The manufacture of this article has been carried on by Messrs. Lynn at Cumberland, very successfully for many years. Mr. Charles P. Manning, chief engineer of the new Baltimore water works, informs me that after having tried all the cements of northern manufacture, he found the Maryland article from the Messrs. Lynn the best. It *sets*, or hardens very quickly, is strong, and impervious to water.

There are six kilns in use, and I was informed by Mr. Lynn that they have machinery which can produce 1,500 bushels per day.

Another cement manufactory is located on the Cumberland and Ohio Canal. three miles above Hancock; the material being also from the lower strata of the limestone, No. 18. It belongs to Mr. Frederick Schaeffer, of Funkstown, Washington county, and is under the management of Mr. Hook. I had an opportunity to witness an experiment in which the cement, upon being properly mixed, became quite hard and strong in a few minutes.

The consumption of cement is rapidly increasing, and will require an increase of the productive capacity of these works and the establishment of others in this limestone, which, it will have been noticed, is abundant in our State. Not only can our own wants be supplied, but the production may be increased to an amount equal to any demand for exportation.

II. BUILDING STONE.

The northwestern half of the State contains many varieties of excellent building stone, which are of local value, and are well known to those who desire to use them in constructions. There are, however, others which, in addition to the extensive local uses, are distributed to the tide water districts, and to some extent exported.

Granite.

There are extensive quarries of granite along and near the Baltimore and Ohio Railroad, at several points between Baltimore and Sykesville. Some of these are actively worked, and give rise to a considerable branch of productive industry.

As the characters of granite were fully described in chapter II, I need only say that the Patapsco granite consists of grains of medium size, the materials being distributed in such manner as to give an uniformly speckled appearance of a gray color. It can be taken out in pieces of any desirable dimensions, free from cracks or seams, and possesses great strength and solidity. The cathedral and record office in Baltimore were built of this kind of granite. There is one variety called porphyritic granite, which has disseminated through its mass reddish colored crystals of felspar, which are supposed by some to improve its appearance for architectural purposes.

Both kinds are equally strong, and are dressed with the chisel when plain or ornamental surfaces are desired.

There is a variety of granite near the Little Patuxent, a short distance above the Baltimore and Washington Railroad, of a *very* light shade of gray, which closely resembles the Quincy granite in appearance.

The Port Deposit granite, as it is usually called, is something between a granite and a gneiss. It has been called syenite by some, but the proportion of hornblende is too small to make that term proper. It occurs in highly inclined strata, and readily separates into flat blocks of convenient dimensions for building purposes, with little or no expense for dressing. This material is used for facing the wharves in Baltimore and other purposes in the tide water counties, and gives rise to a considerable amount of trade.

Sandstone.

The Seneca sandstone is worked at extensive quarries on the Chesapeake and Ohio canal, at the mouth of Seneca creek, in Montgomery county. It was largely used in the constructions on the canal, and also in the public buildings in Washington city. It is easily quarried and dressed, being somewhat soft when first taken out.

There is a quarry of Sandstone at the foot of the Sugar Loaf Mountain, near the mouth of the Monocacy River, which was used for the Canal Aqueduct over that river. It is a firm, solid stone, and is well suited for purposes requiring great strength and durability.

The Brown sandstone so much used for architectural purposes of late has been supplied mainly from New Jersey and Connecticut. Some of it is brought to Baltimore from York county, Pa., by the Northern Central Railroad.

The same formation from which it is procured abroad, (No. 20) on the map, traverses Frederick and Carroll counties. It lies beneath the breccia, before noticed, but crops out from under its eastern borders, and is crossed by the Baltimore and Ohio Railroad, but I believe little has been done towards developing it as a branch of trade. The level character of that part of the county is unfavorable to such operations, because in sinking down for good stone the water soon becomes troublesome.

The north-eastern part of Frederick and the western part of Carroll furnish much greater facilities for procuring this stone. Upon the completion of the Western Maryland Railroad to Pipe Creek there will be a good opportunity (with the aid of low freights) to bring the Maryland Brown Stone into extensive use.

The numerous formations of sandstone, westward of the Monocacy, comprise many varieties of excellent building stone which need not be referred to at this time. They have local uses but will not bear the cost of transportation to distant points.

IV. FLAGGING STONE.

A considerable amount of Sandstone Flagging is brought from New York, and some from along the Schuylkill (above Philadelphia) to Baltimore, and may be seen on many foot-ways in that city. Those in front of the State House in Annapolis are of the mica-slate flags from the Schuylkill.

We have the formations containing them in our State, and if proper measures were taken for opening quarries and bringing these flags to market from divers localities, we might not only supply the home demand but save those South of Maryland from going so far North for their supplies.

The first to be noticed is that near Emmittsburg among the upper layers of the Brown Sandstone of formation No. 20. Some of this separate into flags from two to four inches thick with smooth, straight surfaces. These are likely to come into extensive use upon the completion of the railroad now in progress, which will traverse that formation.

The lower beds of formation 15th consist of fine grained sandstones in thin layers. parted by shales, which furnish the flags so *largely* exported from a precisely similar formation in the interior of New York. In passing westward we first meet it a short distance from Licking Creek, and then again on the Virginia side of the river on Sleepy Creek, two miles from the Baltimore and Ohio Railroad. On Dutch Creek, near Hancock, it presents a very excellent appearance. Two or more ranges of these flags, between Sleepy Creek and Hancock, are crossed both by the railroad and the canal.

There are other ranges crossing the State westward of these which I have not yet been able to examine. Those used in Cumberland are obtained from formation 15th at the eastern base of Dan's Mountain.

V. ROOFING SLATES.

Slates have been long used for covering houses owing to the protection they afford against fire; and large quantities of Welch Slates were formerly imported into Baltimore. Importations of these continue to be made into the Northern ports. The early use of our own slate was retarded by the inferiority of the material obtained near the surface, where it had been acted upon by atmospheric agents and could not be split into thin and large slates. As the quarrying extended to considerable depths a material of the finest quality was reached, so that slates are produced of the largest sizes required, and also very thin. This last is an important point, because thick slates are too heavy for the rafters of ordinary roofs. Those now supplied leave nothing to be desired; they have abundant strength to resist the heaviest hail and wind if properly put on, and yet so thin that a ton of them will cover about four hundred square feet.

The quarries of Roof Slates most extensively carried on in Maryland, hitherto, are in what is called *Slate* Ridge which runs from near Peach Bottom Ferry, on the Susquehanna River, a few miles North of Mason and Dixon's Line. The ridge runs five or six miles west south-west, crossing this line and extends a few miles into Harford county. Quarries have long been worked in the Pennsylvania portion of the ridge, which, besides furnishing supplies for the local demand, have exported it to Baltimore and other cities.

Three quarries have been opened in the Maryland part of the ridge, one of which belongs to Mr. Ludwig, and was not worked at the time of my visit. The second is worked by "The Peach Bottom Slate Minning Company." The third is more extensively carried on by the Messrs. Whiteford. The slates having been penetrated to the depth of about eighty feet are of the best quality and wholly uninjured by atmospheric agencies.

In my travels through the State inquiries have often been made of me for information in reference to the use of slates, their prices, sizes and weight. I therefore collected such facts as would enable me to give definite information upon the subject. As these interest a very large portion of our citizens I will now give them.

The prices at the quarry of the Messrs. Whiteford are as follows:

Length.	Breadth.	Price.
10 inches.	Irregular.	$12 00
12 "	6 to 8 inches.	17 00
14 "	7 to 9 "	18 00
16 "	8 to 10 "	19 00
18 "	9 to 11 "	19 00
20 "	10 to 12 "	19 00
22 "	11 to 13 "	19 00
24 "	12 to 13 "	20 00
24 "	14 to 16 "	$21 to $22 00

It is delivered on board the boats in the tide-water canal, five miles distant, for one dollar additional. The middle sizes, it will be seen, will then cost twenty dollars at the point of shipment, and of these one ton will cover four hundred square feet. The smaller require rather more whilst the larger require less slate to a given area of roof. Small slates of an inferior quality are sold at lower prices.

There is no doubt that intercalations of roofing slate, with some interruptions, extend throughout the length of the talc slate (6) described in chapter III. I have noticed them in Montgomery and Frederick counties, and we may expect that further explorations will develop them in Baltimore, Howard and Carroll.

Near Hyattstown on the borders of Montgomery and, Frederick, a quarry has been opened, and produces excellent slate which is used in the neighborhood, but its proximity to the railroad should enable the proprietor to supply an export demand.

Two quarries of slate are now worked near Ijamsville in Frederick county. They are located upon the Baltimore and Ohio road about five miles eastward of the Monocacy, and fifty-three miles from Baltimore. They produce excellent slate, which will be even better at greater depth.

The prices of the Ijamsville slate as furnished me, are as follows: First quality, five dollars for 580 lbs., which covers 100 square ft.: second quality, four dollars for 620 lbs., which covers 100 square ft.

Slate of good quality is also quarried on the Lingamore in Frederick county, and supplies the local demand for that fine region of country.

Believing that the roofing slates of Maryland may be made a branch of productive industry, and of great moment to both town and country as a protection against fire, it was my desire and intention to have carefully examined the whole range and especially all the quarries now opened. I found it, however, absolutely impossible to do so without still further delaying the preparation of the report which I did my best to hand in on the first day of the session of the assembly.

They will be fully investigated during the next season.

Their importance to Maryland rests upon the following grounds:

1. In their use, as one means of protecting our houses against fire, and for which we are not dependent upon other states or countries.
2. Their value, as an article of export both to the south and west.

By means of our bay, rivers, canals and railroads, this heavy material may be distributed to almost every part of our own as well as many other states. So far as I can ascertain there is no good roof slate known or opened in any southern or western state on this side of the Rocky mountains. The geology of all of them forbids the hope discovering it, unless it be within a very narrow belt, ranging south-eastward through central Virginia and the western parts of North and South Carolina and Georgia, and terminating in the northern part of Alabama. If it should occur along this line there are very few points from which it could be sent, so as to compete with the Maryland slate, either on the sea-board, or in the western or south-western states.

We must bear in mind that the Maryland roof slates are nearer the Great West than any other upon or near railroads. As it is a heavy material, I would suggest to our railroad companies whether it may not be to their interest, as well as that of the public, to charge very low rates of freight on slate, by way of favoring an increased consumption abroad as well as at home.

The present appears to be a favorable time for inaugurating such a system, because of the dissatisfaction expressed by many in regard to the durability of metal roofing. It is more than probable, especially with reference to the exportation of slate to the West, that very low freights (for a time at least,) would stimulate the trader therein, so as largely to increase the trade, to furnish a considerable amount of freight to be taken to the Ohio in return cars that now go empty.

VI. CLAY.

We have many varieties of clay in Maryland, some of which are used, whilst others now neglected are adapted to important branches of industry not yet instituted amongst us. I shall consider each variety of clay separately.

1. *Kaolin or Porcelain Clay.*

This clay differs from all others in being produced by the decomposition of masses of felspar more or less pure, which have existed in or formed a part of masses of granite.

It is from this that the finest kinds of porcelain or china ware are made. It occurs with the granites of Cecil, Harford, Balto., Howard and Montgomery counties, and if the proper manufacturing establishments existed amongst us, inexhaustible supplies of Kaolin can be obtained.

At one point in Harford county, about three miles n. n. e. from Abingdon, I discovered twenty years ago a body of Kaolin of a most superior kind. It is a perfect white and being free from metallic oxides it would furnish porcelain of a pure white color. It is also entirely free from mica, quartz and all other impurities.

2. *Potters' Clay.*

The most extensive deposits of Potters' clays in Maryland are those described in chap. III, as forming the lower members of the cretaceous series in the table at the beginning of that chapter, and in the key to the map.

These clays may be classed with reference to their industrial uses.

1. Pure white fine clay.
2. Those varying in color from dark to light red and sometimes yellow.
3. Lead colored and blue clays, varying from dark to light blue lead color and gray.

The first kind usually cotain but little iron or manganese, and is used in the manufacture of the celebrated Berry's Baltimore fire bricks.

Fire bricks have also been extensively made from these clays in Cecil county, but I believe that the manufacture has been discontinued, because of the prostrated state of iron manufacture, in which they were largely used.

Another application of these clays is in the manufacture of wall paper to which it is *better adapted,* than the more costly whiting from England, formerly used for that purpose.

The same kind of clays exactly, exist in New Jersey, and furnish the material for the manufacture of crockery or delft ware to a very large amount. They produce from it a variety of yellow and mottled crockery ware which have become a considerable amount of trade ; and I am glad to find that some of our own enterprising citizens have entered largely and I doubt not profitably into that branch of industry. So far as I have observed they have not yet found it to their interest to select the clays which burn white so as to imitate the English white crockery.

Such clays exist in this formation, and they will, after a time, come into use and enable us to supply all our wants in this respect.

The lead colored clays in the vicinity of Baltimore have always been celebrated for enabling us to produce two articles of extensive use.

The first consists of building bricks, of which those made near Baltimore are universally considered the finest in the United States for uniformity of color and smoothness.

The second article made of the light colored clays, free from iron or nearly so, is what is termed the *Baltimore stone-ware,* so extensively used throughout the State. It is very strong, and as the glazing is effected without the use of lead or other deleterious matter, and may be used for any purpose for which such ware may be wanted.

The red pottery owes its color to the large quantities of iron in the clay used in its manufacture. The inside of the vessels, and often the exterior also, is glazed with the aid of oxide of lead, and sometimes, oxide of manganese.

Such vessels are wholly unfit to hold any thing containg acids which will act upon the lead. Grave accidents have happened from this cause, and I beg leave to warn the public against the use of this ware to hold any thing whatever containing an acid which is to be eaten or drunk. People have lost their health, and even their lives in this way, and often without the real cause being known until too late.

These clays in every variety, range from Washington northeastward through the State, via Baltimore and Havre-de-Grace to the Delaware line near Elkton. Their aggregate thickness is considerable, and the width of the belt varies from five to ten miles. They are pierced by tide water rivers at numerous points, and the railroads from Baltimore to Washington, as well as to Philadelphia, are almost wholly located in them. All these, therefore, furnish unusual facilities for the transportation of the clays or their products to the points where they may be needed.

Before leaving the subject of clays, I may briefly call attention to the fire-clays of the coal regions of Allegany county. There are several beds of these which have been little examined, except in the Potomac and George's Creek coal field. In this the heaviest beds of fire clay are among the lower beds of the coal formation. At Mt. Savage, near Frostburg, the manufacture of fire-bricks has been extensively prosecuted during many years, and the product are found fully equal to the celebrated Stourbridge (Eng.) bricks. These were extensively used in our State before the Baltimore and the Mount Savage bricks were introduced, and now we export fire-brick and import none.

VII. COAL.

The map shows us that there are three coal fields in Maryland. Of these only the first has been much explored. This has received many names, among which I consider that of the Potomac and George's Creek coal field the most proper, because these, with their affluents, drain the greater part of it. I have made detailed surveys of large portions of this region in former days, and have ample materials to furnish a full account of it, both in its scientific and industrial relations. So much has, however, been published upon the subject that I do not deem it necessary to enlarge upon it at the present time. I may state, however, that the conclusions to which I arrived, and which were published twenty-three years since, have been fully confirmed by subsequent experience in the extensive practical uses of the coal. I then pronounced it the best fuel for raising steam in this country, which is now, I believe, universally admitted by those conversant with the subject.

This formation consists of the following matters:

1. Gray sandstones.
2. Shales.
3. Bituminous shales.
4. Slate clay, or fire-clay.
5. Iron ore, (carbonate of iron.)
6. Coal.

These are interstratified among each other without any regular order of succession. Their aggregate thickness amounts to about 1500 feet.

There are numerous beds of coal, the thickest of which is 14 feet, and there are three others, which are about 6 feet each. The remainder range in thickness from 5 feet down to 1 foot.

The coal at present mined and exported from that region is all taken from the large bed which varies in thickness from 10 to 14 feet. Whilst a portion of the smaller beds contain too much shale to be of value, there are others of excellent quality, and will come into use in after times.

The coal is transported to the seaboard upon the Baltimore and Ohio Railroad and the Chesapeake and Ohio Canal, and is destined to become a vast source of wealth to our State.

West of this we next come to the Meadow Mountain coal field, which has been the subject of very little exploration. I have noticed, near Grantsville, a bed of coal nearly four feet thick, which appeared to be of good quality and free from shale. Another coal, nine feet thick, has been opened, as I am informed, a few miles north of Grantsville, which I have not yet been able to examine.

Leaving the Meadow Mountain coal, we cross Negro Mountain and Keyser and enter the Youghiogheny coal field, which has also been but little explored. One coal bed, about six feet thick, has been opened and mined at Smythfield, a little north of the Maryland line, and it has also been opened for local use at several places in Maryland. The quality of the coal is excellent, and it is free from shale and sulphuret of iron.

Like that in the Meadow Mountain field, this coal contains more bitumen than the coal of the Potomac coal fields. It burns with a clear bright flame and leaves but little ashes.

It is proposed to give a *full* description of these coal fields in a subsequent report.

I cannot conclude this sketch of our coal fields without a word of caution to those who may be disposed to waste money in costly excavations for the purpose of finding coal in the formations beneath what is called the true coal formation, No. 19, in the Table.

Immense sums were expended in this way in New York and Pennsylvania, until the completion of their geological surveys arrested such operations, by showing the true position of the coal and its non-existence in the subjacent formations.

Some years ago about $25,000 were spent by parties in fruitless diggings for coal on the Virginia side of the Potomac river nearly opposite the Licking creek. Large sums have also been expended at Town Hill, Sideling Hill, and at other places in Maryland.

With the exception of *thin*, interrupted seams of anthracite, wholly without value, nothing nearer to coal than a black carbonaceous strata has ever been found in any of our formations lower in the series than the coal formation, No. 19.

As these exist no where in this State eastward of Dan's Mountain, I would say that the expenditure of money in search of it eastward of that line will be attended with more risks of loss than tickets in a lottery.

VIII. IRON ORES.

We are without positive facts as to the precise time when iron smelting was first commenced in this State, but it is certain that iron was imported in England from Maryland in the year 1717. At that period, and for some years after, it was unlawful to make any other than pig iron, which was then smelted exclusively with charcoal. The increasing demand for iron, with the rapid destruction of forests, at length induced the British Parliament *graciously* to permit the establishment of forges for the production of bar iron in the colonies. The act, however, provided that the Americans should not be permitted to erect "rolling mills, slitting mills or forges for making plates, as that would interfere with the manufacturers of Great Britain."

Those who are desirous for further information upon the history of this branch of industry in Maryland are referred to the very interesting report of Dr. Jno. H. Alexander, which was printed by order of the Senate of Maryland in 1840.

I propose to give a passing notice of the more important iron ores within our limits.

Bog Ore.

This variety of iron ore exists in marshes or bogs, wherein its formation is uninterruptedly going on as long as the marsh exists. When drained, however, the deposit of ore ceases.

It usually contains from thirty to thirty-five pt. ct. of the metal in the state of peroxide, and proportions of phosphate of iron, varying from one to fifteen pr. ct., besides earthy and vegetable matters and water.

When the proportion of phosphoric acid exceeds one or two pr. ct., it produces what is called *cold short iron.* It is very readily smelted, and, when mixed in small proportions, promotes the fusibility of other ores, producing a metal suited for smooth and handsome castings.

Bog iron ore abounds in the southern counties of this State, more especially those on the Eastern Shore. It exists in great quantity along the Nasseungo creek, and other affluents of the Pocomoke river. In fact, it is extensively distributed throughout large portions of Worcester and Somerset counties, and also, to some extent, in Caroline and in the eastern part of Dorchester.

It was formerly smelted at Nasseungo furnace, and although pig iron was readily and cheaply produced, yet the brittle character of the metal rendered it unsaleable, and the operation was suspended.

I have not yet had an opportunity to explore that part of the State, but what I saw of these ores a number of years ago, induced the belief that some of them contain very large proportions of phosphoric acid, whilst from others this substance was nearly or altogether absent.

Although so injurious to the iron smelter, phosphoric acid, as we have seen, is an absolute necessity to the former, and I consider it a duty to do all in my power to obtain it at the lowest rates, and, if possible, within our own State. When I made an attempt at a special examination of these ores, with reference to their phosphoric acid, both in the ore and in the subjacent clay, I was informed, by those familiar with the ore deposits, that the water was too high for my purposes. I was promised to have sent me a barrel of the ore from two of the largest deposits, but they did not come to hand. I propose to take an early opportunity to investigate them in all their relations.

Circumstances which have of late come to light, give an increased importance to these deposits. Phosphate of iron, as I have before stated, has been proven to be soluble in alcaline silicates, which must exist in every soil in which grass or grain can grow. It has also been proven that oxides of iron whilst undergoing certain changes in the soil promote the formation of ammonia and carbonic-acid. This will probably account for the fact of the growth of fine tobacco in a soil containing 90 pr. ct. of iron as related in chap. XVI.

2. *Brown Hematite.*

This species of iron ore presents several varieties which occur in different parts of the state. Their chemical composition is that of a peroxide of iron united with about 15 pr. ct. of water. When free from earthy impurities it contains nearly 61 pr. ct. of iron, but it always contains more or less of earthy matters, so that as used in the furnaces it usually yields from 35 to 45 pr. ct. of metal. One locality of this ore has come under my notice, in Anne Arundel county, about two miles south of Owingsville.

It exists also in considerable quantities near the edges of the metamorphic limestones in Baltimore county, and is smelted on a large scale at the Ashland and Oregon furnaces near Cockeysville.

In Carroll connty it occurs in *immense quantity* in connection with the limestones (11) before mentioned. They range from the Pennsylvania line (north of Westminster) southwesterly for ten or twelve miles. Westminster lies on the eastern edge of the range. There are the ruins of an iron furnace about 2½ miles southwest of Westminster, on the property of Mr. Vanbibber, where these ores were smelted many years ago. The Western Maryland railroad will reach this range of ore at Westminster and pass through it for several miles. This will afford every facility for transporting the ore or the iron that may be made therefrom.

There are also hematitic ores still further southwest on Parr's ridge, which I have not yet examined.

Near the western base of the Catoctin mountain there are very important and extensive deposits of this kind of ore, from which the supplies for the Catoctic furnace have been drawn for about 85 years, and which is now carried on by Col. John M. Kunkle. The ore near the Point of Rocks supplied one of the furnaces of Governor Johnson during the revolutionary war. The same range extends along the Catoctin into Virginia.

On the west side of the South mountain near its base, there are other varieties of hematite to which *pipe* ore and other local names have been given. Those near the Potomac were long used at the Antietam furnace, and the iron therefrom always possessed a high reputation for its strength. The ore occurs at many points along the base of the mountain. It occurs again at the base of the North mountain near the Potomac river, and is now smelted at the furnace of J. Dixon Roman, Esq., near to old Fort Frederick, and upon the site of a furnace erected before the revolutionary war.

Westward of this, it is believed there are no important beds of hematitie, but there are ample supplies of other iron ores which will be spoken of presently.

3. *Specular iron ore*—(Red oxide of iron.)

The ores of this species present themselves under different aspects, but they may be very easily distinguished from other ores, from the fact that when powdered the color is always dark red, whilst that of bog ore and hematite is yellowish brown. Some varieties which have a crystalline or a compact structure with a metallic lustre, resemble the magnetic iron ore: but the powder of the latter is nearly black and it is attracted by the magnet which has no effect upon the specular and red oxides.

When pure, this ore contains 70 pr. ct. of iron, but it is always more or less mixed up with earthy matters. The richest ore of this kind in Maryland is in the eastern part of Frederick county, where it constitutes one of the minerals of the metalliferous district between the Monocacy and Parr's ridge, which will come under our notice in treating of the copper ores of that region. I have noticed pieces of very rich ore lying upon the surface south of Libetry, but I believe no attempts have been made to ascertain whether it can be had in available quantities. And the same may be said of the specular ores in the Catoctin mountain.

One of the most important iron ores and which supplies a large proportion of the Pennsylvania furnaces, occurs in the surgent shales (14 a) in the table. These pass through Maryland on the western slope of the north mountain, and again in narrow belts along several lines further west, but without furnishing available quantities of ore, until we reach Sideling hill and Townhill. Little, however, has been done in developiug the ore except at Wills mountain, on both sides of which it has been extensively mined for the Mount Savage iron works. It extends southwestward into Virginia, and has been mined for the use of the Lonaconing and Mt. Savage furnaces.

In this formation it is interstratified between the calcareous shales containing fossil shells, which has caused the distinctive name of fossil ore to be given to it. Some of its layers contain phosphoric acid in sufficient proportion to affect injuriously the quality of the metal. The general color of the fossil ore is reddish and it is without metallic lustre, but glistening scales of the specular oxide are not uncommon.

In the lower part of the formation we have what is called the "Hard ore" containing from 25 to 30 pr. ct. of iron mixed with grains of sand. This is too silicious to be smelted without being mixed with other kinds of ore.

4. *Magnetic oxide of iron.*

This is the richest ore of iron, and when pure (as it is sometimes the case in Sweden,) contains seventy-two per ct. of metal. It is usually however, more or less mixed up with earthy matters, and sometimes contains the oxides of titanium and manganese.

It has a metallic lustre and a dark grey or almost black color, the latter being also the color of its powder. It strongly attracts the magnetic needle, and when in small grains it is attracted by the magnet. Some of its varieties are sufficiently magnetic to attract iron filings or a small needle; hence the name of loadstone which was formerly applied to it.

These characters distinguish it from all other ores of iron.

It occurs in small quantities about seven miles west north-west from Baltimore, near the Barehill's Copper Mine, and again near Scott's Mills, about eighteen miles north, north-west from Baltimore. The only important localities in Maryland are in the metalliferous range in north-western edge of the mica-slates (formation No. 5) in which it is usually associated with the chlorite slate noticed in chapter III. It occurs massive as well as in octahedral crystals and grains. An Iron Furnaee at Sykesville, in Carroll county, is in part supplied by ore which is mined in that vicinity.

It is also obtained in Baltimore county, near the Northern Central Rrailroad, about twenty-four miles from Baltimore, and is used at the Ashland Furnaces, near Cockeysville. Near Dear Creek, in Harford county, a variety was mined and used at the Harford Furnace near Bush. It contained 18 per cent. of oxide of titanium.

5. *Carbonate of Iron.*

In chapter III some notice was taken of what I have provisionally termed *Iron Ore Clays*, a portion of the cretaceous formation. They are numbered 22 in the table, and their position is shown to be included within the north-western portions of formation 21.

There are some very interesting geological considerations connected with these clays which were alluded to briefly in chapter III and will be duly attended to in a subsequent report.

The iron ore in these clays attracted attention at an early period in our history and has always been celebrated for the excellence of its iron.

The ore occurs usually in flattened nodules weighing from 100 lbs. downwards. In some localities there are irregular shaped masses of considerable size. It has a close, compact structure and a grayish dun color which has caused the name of *hone ore* to be applied to it.

It contains from 33 to 40 per cent. of metal *combined* with oxygen and carbonic acid, and more or less mixed with earthy matters and small proportions of manganese.

The Pig Iron produced from this ore, when smelted with charcoal, has always maintained a high reputation, and when made into bars it is especially esteemed for all purposes wherein strength and toughness is required. The high price of wood has induced many iron masters to smelt this this ore by means of Anthracite Coal, by which excellent foundry iron is produced.

There are a number of furnaces existing along this range which extends from near Elkton, through Baltimore to the vicinity of Washington. Several of these are in operation, but most of them are "out of blast" owing to the depressed condition of the trade in this country at this time.

Westward of the ores in this group of clays, we do not again meet with Carbonate of Iron until we reach the shales numbered 14*a* in this table. The first range of these (in the lower part of which we may expect the ore) is between Licking and Tonaloway Creeks. I believe, however, they have not yet been discovered therein. Out-crops of ore appear in these shales in Sideling Hill and Town Hill. These will doubtless prove to be the carbonate if they be penetrated beyond the reach of the oxydating effects of the air, which changes the carbonate into an ore analagous to hematite.

Near the mouth of Town Creek a thick stratum of an ore of this kind exists, which, from its proximity to the canal and railroad as well as to the coal region, cannot fail to be made available upon the revival of the iron trade.

As we find in Pennsylvania in the umbral shales, 18*b*, an important bed of Carbonate of Iron we may expect it in corresponding geological positions in Maryland. On Dunkard's Creek, near Uniontown, Fayette county, Pa., it occurs in a continous stratum varying from one to three feet thick. Its position is a little above the limestone, as shown in the map, whose extensive outcrops in Alleghany county were described in chapter III. It may be confidently expected that further explorations in bringing this ore to light will materially add to our facilities for *cheaply* manufacturing iron.

It is used at Union Furnace, on Dunbar's Creek, and I was informed by the proprietor that one ton of iron is produced from two-and-a-half tons of roasted ore which costs $2.50 per ton, or $6.25 for the quantity required for one ton of iron.

The Carbonate of Iron in the Coal basins are next to be noticed. They occur in the three coal fields in Maryland, but have been little explored, except in that nearest Cumberland which is accessible by the Valleys of Jenning's Run, Braddock's Run, George's Creek and the Potomac River. I have abundant material for a full account of this coal field and its ores, but the plan of the present report does not permit me to embrace these details.

I may state, however, that the carbonate of iron of the coal regions consists of

1. Flattened nodules, called *balls* by the miners. These are embedded in shales and sometimes in fire-clays.

2. Regular strata, varying in thickness from an inch or two to several feet. These are intercalated either between beds of shale or of coal.

I have a complete section made from actual measurement of the whole thickness of the Potomac Coal Field, showing the position and thickness of each bed of sandstone shale, coal, limestone, iron ore, &c. This was made whilst professionally engaged in surveys some years since and will be availed of in a subsequent report.

Whilst many of the strata contain ores in too small quantity to be profitably mined, there are others possessing a high industrial value. One of these, about one hundred and fifty feet below the main coal, contains three courses of nodules and one band of ore, which have been extensively used, and proved to produce about 33 per cent. of foundry iron of the best quality.

About forty feet below this is a very important stratum of ore of the variety called *black band*, so much prized in England and Scotland, The result of very careful surveys which I made in that region several years ago indicated that this band of ore will prove highly valuable for iron manufacture. It rests upon a seam of coal three inches thick which will enable the miner to take it out for seventy-five cents per ton. As mined it contains a sufficient amount of coal to roast it, when the fire has been started.

It requires 3⅓ tons of raw ore or 2½ of roasted ore to produce one ton of iron.

The only points at which it was developed were on Koontzs Run and Mill Run, which head in the Savage Mountain and flow into George's Creek.

At no great distance below this I found upon Koontz Run a solid stratum of ore, five feet thick, containing 25 per cent. of iron. Still lower in the series are numerous strata containing both bands and balls of very superior ore which have not yet been brought into use.

It is from ores of this kind that nine-tenths of the iron of Great Britain is made, and a good deal is also made from them in Pennsylvania and Ohio. The day is probably not distant when they will be largely used to increase the products of iron in Maryland.

I have as yet had no opportunity to examine the ores of the Meadow Mountain, or of the Youghiogheny coal field. From the latter I have specimens of the richest ores of this kind.

Iron Pyrites.—(Sulphuret of Iron.)

This material occurs at several points in the cretaceous clays, (No. 21.) At Cape Sable, on the Magothy river, in Anne Arundel county, it was formerly used as a material for the manufacture of alum and copperas. It also occurs, in apparently large quantity, near Oxon creek, in Prince George's county, south of Washington. An improvement, called the *Monier process*, has been recently brought into use, by which sulphuric acid is readily produced from iron pyrites more cheaply than from sulphur. It is not improbable that the pyrites in this range will be brought into use for this purpose.

ORES OF COPPER, LEAD, ZINC, CHROME, MANGANESE AND GOLD.

There are certain portions of our State containing copper and other ores, (exclusive of *deposits* of iron ores,) and which are termed metalliferous districts. Some of these occupy small isolated areas, whilst others constitute long ranges.

One of these, consisting of a thin seam in gneiss, containing galena or sulphuret of lead and sulphuret of zinc, was discovered in one of the quarries on Jones' Falls, near Baltimore, many years ago, but was never explored. It is probable that it was not worth working.

A serpentine formation, about seven miles N. N. West from Baltimore, contains chrome in the state of an oxide combined with oxide of iron. The district is called the Bare Hills, because of its sterility, owing to the large quantity of magnesia in the soil.

It was there that chrome ore was first obtained in sufficient quantity to be manufactured into a paint. It is still a very rare mineral in Europe, except at a few points in Russia and in Turkey, from whence it is transported to England, France and Germany, so that our people now find rivals in the trade. A few years since the world was supplied exclusiveiy from Maryland and from points in Pennsylvania adjacent to Cecil county.

A copper mine was opened near the eastern border of the Barehills, and, after being worked a few years, the mining was suspended. A good deal of ore was taken out, and it was certainly a promising mine when I saw it before the working ceased. It has recently changed hands, and will doubtless be fully examined and worked if it shall prove productive, which is highly probable.

Southwestward of the Barehills, about six miles, is another formation of serpentine, called Soldier's Delight, from which chrome ore, as well as a magnesian mineral, used for making Epsom salts, was formerly obtained.

When we reach the northwestern edge of the mica slates, (No. 5) we find what may be termed a metalliferous range, extending from the northern part of Cecil county, through Harford, Baltimore, Carroll, Howard and Mont gomery counties.

In addition to the magnetic iron ores of this range already referred to, there are ores of copper, chrome, manganese, and gold.

Indications of copper may be seen at various points, and several mines have been opened in Carroll county, one of which, at Springfield, near Sykesville, continues to be efficiently carried on.

Near Finksburg, in Carroll county, a copper mine was successfully worked during several years, and, when I examined it, was certainly quite promising. When I passed it, a short time since, it was idle, but I am strongly inclined to the opinion that if the proper skill and sufficient capital were applied, it will prove productive. The ore consisted of yellow or pyritous copper, and one still richer called purple copper ore.

I first discovered sulphuret of cobalt among the products of this mine, but this rare and valuable material occurred in very small quantity, and has not been found elsewhere in this State.

Other mines were opened in this range, between Finksburg and Sykesville, but I believe they are not in operation at the present time. At one of them native gold was discovered.

Northeastward from Finksburg there are indications of copper at many points, especially near the forks of the Gunpowder river, about twenty-two miles north from Baltimore. Some explorations and diggings have been made without discovering the ore in quantity. It appears to be associated with the magnetic oxide of iron of this formation.

The map shows several formations of serpentine, near the northwestern borders of this range, and it is in connection with these that chrome ore is always found. There are, however, but two localities wherein it has been extensively mined, although it has been attempted in numerous places. A mine was worked with large profits for many years near Cooptown, in Harford county, and then suspended because of the discovery of still more productive mines in Cecil county, and the adjacent parts of Lancaster county, Pennsylvania.

Harford and Cecil county have largely contributed the magnesian mineral from which nearly all the Epsom salts, sold in this country, is made.

I have had little opportunity to examine this range in Montgomery county- We know, however, that good oxide of manganese exists therein, but the work. iug did not prove profitable. Gold also occurs in that county, and a mine was opened and worked some ten years since, but, as is the case with most gold mines, the pure metal was found to cost more than it was worth.

The eastern face of Parr's ridge seems destitute of metals, except iron ores in no great supply; but, upon crossing the summit, we find a very interesting metalliferous district east of the Monocacy. Indications of copper are abundant at many points between the Baltimore and Ohio Railroad and the Pennsylvania line, and generally in connection with the isolated areas of limestone in that region. We may designate it as the Linganore copper region, because that stream and its affluents drain the greater part of it.

Copper ore was mined and smelted near Liberty before or during the Revolutionary War. About thirty-eight years ago an attempt was again made, and persevered in for some time, to open a productive mine, but it was at length discontinued.

A few years since a mine was opened adjacent to the town of Liberty, on Dolohyde creek, and for some time was very productive, but, becoming less so, the works were suspended. Other attempts have been made in this district with like results.

I have had occasion, in former years, to make special examinations of portions of this district, and have come to the conclusion that there is a fair probability it will yet be found to contain valuable mines of copper. There are some intricacies in its geological structure which it would altogether exceed my limits to enter into at this time. A brief reference to the subject, however, seems proper.

Copper and some other metals occur under the following conditions:

1. In veins, which are cracks formed in rocks and afterwards filled by deposits from solutions aided by electric agencies.

2. In some cases they have been forced up from below in a state of fusion like intrusive rocks.

3. There are also cases in which intrusions of rocks have taken place, and contracting whilst cooling, have left cavities which afterwards were filled in the same manner as veins.

All European mines, except a portion of those in Sweden and in Tuscany, (including the island of Elba,) are in veins; so that the European mining captains who come here, have learned their geology from their experience in that kind of mining. Veins are continuous in nearly straight lines, sometimes for several miles, whilst a *contact mass* which occur in the second and third cases *never* extend to a great distance, but often contain immense deposits of iron or copper as in Cuba, Chili, Missouri and Michigan and other regions.

When I saw the Dolohyde mine in operation, the captain (an Englishman), thought he was working a vein, and of course wasted money by endeavoring to trace its prolongations. I told the parties by whom I was consulted, that it was a *contact mass*, and that the ore was to be sought for by working in the outer edge of the isolated mass of limestone (with which it was associated), and by resolutely sinking downwards. I also stated that no reliance should be placed in the horizontal extension of the *false* vein they were then working, which, although it contained a large proportion of rich ore, would not extend horizontally 100 yards. And such was the result.

In Tuscany, there are a few copper veins which were skillfully mined by the Romans more than 2000 years ago. The best of them were exhausted long since. The Tuscans have also *contact masses* which miners had often attempted to work, but failed to do so with profit. The *contact masses* being thin, irregular and unpromising at the surface, were much neglected. At length a French geologist (I think it was M. Burat), studied them, and by his advice one of them was penetrated downward, which brought to view extremely rich ores of copper in enormous masses.

Galena, a rich ore of lead containing silver was obtained in the Dolohyde mine, and had previously been discovered near Unionsville. I have also heard of its occurrence at other points, but have no definite information in reference to them.

It is my purpose to examine this Linganore copper region in Frederick county, with all the care that its importance demands, and endeavor to propose some system by which this buried wealth may be made available.

We have also abundant traces of copper in the red shales (No. 20) in the n. e. part of Frederick, and the n. w. part of Carroll county, and so there are in the same formations in other states. They give, however, so little promise of profitable mines, that I would not advise the expenditure of money in digging for the ore. Much has been spent in searching for copper ore in those shales without useful results.

We have another metalliferous region, embracing the Catoctin mountain and Middletown valley, about which little is yet known. Fine specimens of pure native or metallic copper have frequently been found in the mountain, as well as stains of carbonate of copper upon the slates both in the mountain and the valley. A citizen of Frederick county also informed me of the existence of calamine, (an ore of zinc) in the mountain, but I have not seen even a specimen of it.

This metalliferous range also deserves to be most carefully investigated.

ARTESIAN WELLS.

Many years ago, I was fully satisfied that Artesian wells were suited to a large portion of our state. In the year 1828, I prepared an article upon the subject for the American farmer, at the request of the late John S. Skinner, its first editor. It was illustrated by a diagram to aid in explaining the cause of the rise of water in these wells. The object was, to call the attention of our people to this means of obtaining water in districts where there are few or no springs affording good water, and where it could not be obtained from wells of the ordinary kind and depth.

The subject was neglected in a great measure, owing to the fact that we were without operators to sink wells of that kind, until about seven years ago the business was commenced here by Mr. J. N. Bolles. Subsequently other parties also engaged in it. They all, I understand, use the patent flush pipe invented by Mr. Bolles, which avoids the difficulties in sinking formerly experienced.

I am informed that about 100 Artesian wells have been sunk in Maryland, nearly all of which are in Baltimore and its vicinity. Of these, it is stated, that about 90 have proved completely successful in bringing good water above or near the surface.

It was my intention to have added to the present report a full account of Artesian wells, accompanied with the proper drawings, but the delay in the preparation of the report from the causes before stated, makes it necessary that I should touch very briefly upon this subject.

There is no time to prepare the drawings and get them engraved and I do not feel justified in delaying the printer. I must defer a full account for another occasion.

The rise of water in Artesian wells to or near the surface, depends altogether upon the geological structure of the district in which they are sunk.

They succeed best where there are alternations in strata of clays and sands or gravel, or where they consist of porous sandstones and slates.

In intrusive rocks, such as from one to four, they so very rarely succeed that I would advise against attempting them in any case.

In metamorphic rocks (5, 6 and 7) it is a very rare case, indeed, that as much water can be obtained from an Artesian well of considerable depth as from the ordinary wells of the country.

In boring deep into limestones there is a great uncertainty as to the supply of water, owing to the caverns or seams in this kind of rock.

The disposition of the strata is to be carefully considered. If it be basin or trough-shaped and the well be sunk through stratified rocks, the water *will* rise to or above the surface. If the strata for many miles on each side of the well dip in the same direction success is very uncertain. Upon an anticlinal axis it is still more hazardous.

Having laid down these few principles, I shall ask the reader to cast his eye upon the left hand side of the map and the lowest section in the "Illustrations," whilst I point out the characteristics of each district as we pass on to the right or eastward.

First is the Youghiogheny coal *basin* in which the water in the Artesian wells will doubtless rise to the surface and as a general rule the nearer the axis the higher it will rise. The same may be said of the other two coal basins, but in the intervening Old Red sandstones and shales there would be little chance for water rising to the surface because of their stratification being anticlinal (the reverse of basin or trough-shaped.) And so it is with Wills' Mount; but in the trough-shaped strata between that and Evitt's Mount, a little east of Cumberland, we have another favorable district for our purpose. From thence to Tonoloway Hill the strata are somewhat disturbed and have not yet been sufficiently explored to enable me to point out with certainty all the favorable districts for these wells. It is most likely they would answer between Evitt's and Martin's Mountains and probably between Town Hill and Sideling Hill.

Between Tonoloway Hill and the North Mountain we have a wide district, whose trough-shaped strata are highly favorable for Artesian wells, especially towards the middle of it.

As the great valley of Washington county, between the North and South Mountains, is principally underlaid by limestone, the details of whose stratification have not yet been made out, it is not possible to form correct opinions relative to the success of such wells. The survey proposed in chapter VIII, for the purpose of determining the position and qualities of the limestones best suited for agricultural lime, will also throw much light upon the adaptation of these strata to Artesian wells.

There is some uncertainty as to Wells in Middletown Valley, and besides the boring would be very costly because of the Trap and other hard rocks that are to be encountered in many parts of this valley.

The New Red Sandstone formation of the Monocacy Valley it will be observed dips to the West, but there is no doubt that its western edges have been tilled up. So far, however, the strata appear to be concealed by the detritus from the Catoctin Mountain. The trough thus formed presents a favorable district for Artesian Wells. A well of this kind has been sunk in a ravine of the Catoctin to aid in supplying the City of Frederick with water. I have not yet been able to collect the facts relating to it, owing to the absence from that city of the gentleman who had charge of it. I learned, however, that it was sunk in formation No. 8, westward of the New Red Sandstone No. 20. A much larger supply would have been obtained if the well had been located in the latter, but it would have been at too low a level to be conducted by a natural flow to the city.

Within the formations between the Monocacy valley and the tide water districts, or rather between No. 10 and No. 21, the circumstances are unfavorable to sinking artesian wells. The few attempts that I have heard of in this region have been failures, whilst the cost of drilling through the hard rocks was enormous.

It is in the cretaceous clays, No. 21, that by far the larger portion of all the wells of this kind in Maryland have hitherto been sunk. In depth they range from forty to one hundred and eighty-eight feet, but those in the lower parts of Baltimore will average about seventy feet. Many of them reach the gneiss and other rocks, which underline formation 21. Where the water does not flow above the surface it is so near as to be pumped out very readily.

The tube or pipe is usually eight inches in diameter, and as it completely shuts out all the surface and other impure waters from the upper strata, the water from these wells is, in almost every case, equal to the best springs of rock water. The only exception is where the parties finding an abundance of water at certain depths that answered their special purposes, did not wish to incur the expense of going deeper for a purer article.

About seven miles southeast of Baltimore there are three artesian wells. One is at the head of Bear creek, on the north side of the Patapsco, and another at the head of Curtis' creek, on the Anne Arundel side. Both of these furnish most copious supplies of the finest water for drinking and culinary purposes, from a depth of one hundred and seventy-eight feet.

A well has been sunk to the depth of one hundred and sixty feet at Fort Carroll, now being constructed in the Patapsco river, between the other two wells. When I was last there the water was strongly chalybeate, but as the strata consist of alternations of sand, clay and gravel, they have only to penetrate below a thick bed of clay to shut out the chalybeate water. They may expect to find a stratum of sand or gravel, containing good water, at a depth not exceeding two hundred feet. The tube in this well has a diameter of twenty inches, and it would long since have been finished but for a sort of small economy on the part of the General Government, which has several times arrested the work.

An artesian well was sunk at Annapolis some years since, within the grounds of the Naval School. The diameter of the tube is eight inches and the depth two hundred and twenty feet. The water discharged seven feet above the surface is at the rate of eleven thousand gallons in twenty-four hours, except when the pipe becomes choaked up with sand. This happens because the sinking was improperly stopped in a loose sand, instead of being continued down to a firmer stratum of sand or gravel. The strata pierced by this well belong to the cretaceous or green sand, and the water is chalybeate.

I have before had occasion to allude to the artesian well proposed to have been sunk in the State House yard at Annapolis, two years ago, but which was omitted because of the insufficiency of the appropriation to improve the State House. If it be the pleasure of the Assembly to provide an ample supply of water for drinking, as well as for the protection of the venerable State House from fire, I beg leave to make the following suggestions:

1st. That the appropriation be made specific for this purpose.

2d. We cannot say at what depth precisely we shall meet a stratum of firm sand giving pure water; but, to avoid the errors of the General Government at Fort Carroll and at the Naval School, it would be safer to estimate going down eighty feet lower than the latter. Add to this the difference in level, say fifty feet, we would have three hundred and fifty feet as a depth *within* which we are sure of a full supply. This, at $4.50 per foot, will amount to $1,575; but as other fixtures would be required for distributing the water and for getting rid of the waste, the appropriation should be for $2,000.

A successful well of this kind would not only be invaluable for the purposes before stated, but would furnish information relating to such wells of great value to our citizens in several tide water counties. By preserving an accurate account of the strata passed through, we should be materially aided in determining the probable depth of wells required to supply good at or near the surface, in a number of those counties.

We all know the necessity of good water to preserve health; and that this has been effected by means of such wells there is no doubt whatever. I shall give an illustration: The low ground, (but little elevated above tide-water) near the head of Bear creek, was long noted for its insalubrity. Intermittents and other diseases prevailed, especially late in summer and in the autumn, because of the water of their shallow wells being loaded with organic matters during those seasons. During the winter, spring and early summer, the larger supply of water afforded to these wells dilutes these poisonous solutions. It is only when they become more concentrated, later in the season, that their baneful effects upon man, and doubtless upon other animals, is fully experienced.

The remedy for all this is to use water, from depths at which it is found, free from these deleterious matters. I am informed that since this was done by means of the well on Bear creek, intermittent fevers have been altogether banished. It has long been the opinion of some eminent medical men that much more disease is produced by drinking impure water than by inhaling what is called miasma. Chemistry has utterly failed to detect this *imaginary* substance in the air of malarious districts; but it shows us that late in summer the water in shallow wells in low districts of country abounds with unwholesome matters.

A well was commenced by my friend, Dr. R. S. Stewart, at Dodon, in Anne Arundel county, but, owing, I presume, to a want of skill in the contractor, "it got a twist," as they call it, at a considerable depth, which, of course, arrested further operations. This is much to be regretted, on account of its enterprizing proprietor, as well as the public. But for this untoward result, the well would have been continued, so as to bring to or near the surface an abundance of good water, and it would have given information highly useful to those residing upon the same geological formations.

The present state of our knowledge leads to the conclusion that there is a general southeasterly dip in our formations as indicated in the sections. Consequently artesian wells must be deeper in proportion as we proceed southward and eastward. It is to be hoped that an important addition to our knowledge in this regard, will be furnished by the completion of the 8 inch well at Centreville, in Queen Anne's county.

I regret that here as at the Naval School, an accurate record of the strata was not preserved. I am promised, however, that when the boring shall be resumed, such a record as well as samples of the borings will be furnished me.

From what I saw of these remaining, although much mixed together, I concluded that the cretaceous green sand or Jersey marl had been reached, and penetrated for some depth. This rarely produces good water which, however, we are certain to reach in the sands and clays beneath.

We are yet without information in reference to Artesian wells in the southern counties on both shores. Private individuals are unwilling to attempt such wells with a prospect of being obliged to sink them to a considerable depth. If each of these counties will follow the example of Queen Anne's and sink *effective* wells in their county towns, we should collect a map of facts of great value to all the tide water counties. The cost of sinking artesian wells in the west, through limestones, sandstone shales and other rocks, in search of salt water, appears to be less than is charged for those in our clays and sands. Drilling is necessary in rock, but in the clays the boring tool is in the shape of a large corkscrew which fits inside the pipe or tube. The screw enters the beds by being turned by men or by steam, and the tube is forced down with the aid of heavy levers. If the higher charges here be owing to less competition among contractors, the difficulty will soon cure itself.